Basic Concepts in Statistics

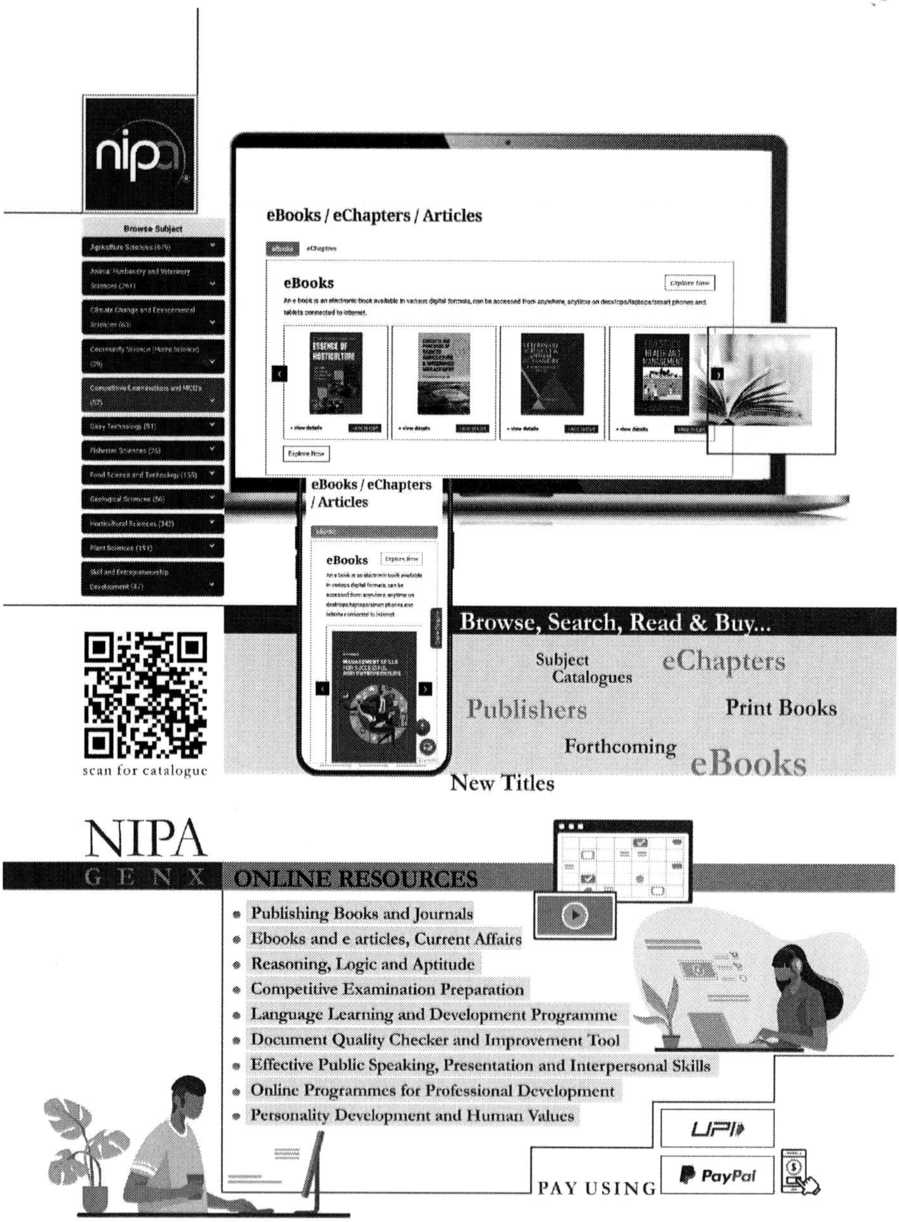

nipa
Browse Subject
Agriculture Sciences (679)
Animal Husbandry and Veterinary Sciences (261)
Climate Change and Environmental Sciences (63)
Community Science (Home Science) (39)
Competitive Examinations and MCQ's (57)
Dairy Technology (91)
Fisheries Sciences (26)
Food Science and Technology (155)
Geological Sciences (56)
Horticultural Sciences (342)
Plant Sciences (191)
Skill and Entrepreneurship Development (47)
eBooks / eChapters / Articles
eBooks
Explore Now
An e book is an electronic book available in various digital formats, can be accessed from anywhere, anytime on desktops/laptops/smart phones and tablets connected to internet.
Essence of Horticulture
Explore Now
eBooks / eChapters / Articles
eBooks
Explore Now
Management Skills for Successful Agri Entrepreneurs
scan for catalogue
Browse, Search, Read & Buy...
Subject Catalogues
eChapters
Publishers
Print Books
Forthcoming
eBooks
New Titles
NIPA
GENX
ONLINE RESOURCES
Publishing Books and Journals
Ebooks and e articles, Current Affairs
Reasoning, Logic and Aptitude
Competitive Examination Preparation
Language Learning and Development Programme
Document Quality Checker and Improvement Tool
Effective Public Speaking, Presentation and Interpersonal Skills
Online Programmes for Professional Development
Personality Development and Human Values
UPI
PAY USING
PayPal

Basic Concepts in Statistics

Authors

K.S. KUSHWAHA

M. Sc., P.S.C.C., Ph.D. (Statistics)
Recipient of Dr. Radhakrishanan Award (1992)
Associate Professor (Statistics)
Department of Mathematics & Statistics
J. N. K. V. V., Jabalpur, M. P. India

&

RAJESH KUMAR

M. Sc, P.S.C.C., Ph.D. (Statistics)
Principal Scientist (Statistics)
I.I.S.R., Lucknow, U.P. India

New India Publishing Agency

Pitam Pura, New Delhi- 110 088

Published by
Sumit Pal Jain *for*
New India Publishing Agency
101, Vikas Surya Plaza, CU Block, L.S.C. Mkt.,
Pitam Pura, New Delhi- 110 088, (India)
Phone: 011-27341717, Fax: 011-27341616
E-mail: newindiapublishingagency@gmail.com
Web: www.bookfactoryindia.com

Preface

The book entitled "Basic Concepts in Statistics" has been written for all P.G., Ph.D. students in the statistics, social science, biological science and faculty members of statistics, agricultural statistics, statistics, agricultural scientists and other students who have to appear in competitive examinations with statistics as a subject in state P.S.C., U.P.S.C., A.S.R.B. & I.S.S. etc. This book is the outcome of 24 years of teaching experience to U.G., P.G. & Ph.D. students of the disciplines, agricultural engineering and agricultural statistics.

The book contains 15 chapters covering different topics of statistics and has been written with views for its readers and users to get familiar with basic knowledge of important topics of statistics. This book will broaden the horizon of basic concepts and knowledge of its readers.

During the preparation of the manuscript, the author has incorporated the fruitful academic suggestions provided by Dr. Housal P. Singh, (Prof. & Head, School of Studies in Statistics, Vikram University, Ujjain, M.P.) to enrich the book.

The authors are very much thankful and extends their honourable honor of high order to the Vice Chancellor (JNKVV, Jabalpur) and Director (IISR, Lucknow) for their kind permission to hand over the manuscript to "New India Publishing Agency" New Delhi to publish it in the shape of a book.

The author extends his elderly affectionate and caring attitude to a very young man, Mr. Manoj Kr. Maurya for taking keen interest in providing physical facilities to get this manuscript back from Yojana Bhavan, U.P. Units of Govt. of India and also to a very young and distinguished Dr. Sehar of K.G.M.U. Lucknow for providing moral support to the author for getting rid of his bad health condition during (Nov. 2006-Feb. 2007).

The author acknowledges very strongly his wife Smt.Veena Kushwaha, daughter Miss Akansha Kushwaha, son Mr. Utkarsh Kushwaha and sister-in-law, Mrs. Seema Maurya who cared the author a lot during the preparation of this manuscript and his bad health.

Last but not the least; the author is also very much thankful to the entire staff of New India Publishing Agency, New Delhi for taking keen interest in publishing the manuscript in its present shape.

Although, the authors have tried their best in the formulation of the subject matter very nicely but that can be further improved based on the suggestions received from its readers. Therefore, the constructive and academic suggestions received from all sides will always be welcomed and highly appreciated by the authors.

June, 2008
Jabalpur

K.S. Kushwaha
Rajesh Kumar

Contents

SECTION-B : SAMPLING DISTRIBUTIONS AND ASSOCIATED TESTS OF SIGNIFICANCE

SECTION-C : STATISTICAL INFERENCE

SECTION-A
DESIGN OF EXPERIMENTS

Chapter 1

Preliminaries

Q.1. What is an experiment ?

Answer

An experiment is a means of getting an answer to the question that an experimentor (researcher) has in mind. This may be to decide which of the several pain relieving drugs that are available in the market, is the most effective or whether they are equqlly effective.

An experiment may be planned to compare the yield of different varieties of wheat. so that the varity giving the maximum yield may be recommended to the farmers to increase their production.

Q.2. What is design of experiment ?

Answer

Designing of an experiment means decidng how the abservations or measurements should be taken on experimetal units to answer a particular question in a valid, efficient and economic way.

Remark : The design and final analysis go together. They are inseparable in the sense that if an experiment is properly designed then there will exist an appropriate way of analysing the data. From an ill designed experiment, no conclusion can be drawn.

Q.3. What are the two categories in which the design of an experiment may be classified ?

Answer

The following are the two categories in which the design of an experiment may be classified

(i) Absolute experiment

(ii) Comperative experiment

Q.4. Define absolute and comperative experiment?

Answer

Absolute Experiment : An experiment through which an absolute value of some characteristic under consideration is determined, is known as an absolute experiment. For examples *(i)* obtaining the average intelligence quotient (I. Q.) of a group of students. *(ii)* Finding Correlation coefficient between two variables in a bi-variate population.

Comperative Experiments : An experiment through which the effect of two or more treatments under consideration on some population characteristic are compared, is known an comperative experiment. For examples *(i)* comparision of different manures or fertilisers *(ii)* Varities of crops etc.

Q.5. What is an experimental units ?

Answer

The smallest division of the experimental material to which the treatments under consideration are applied and observations are taken, is known as experimental unit. For examples *(i)* In the field experiments, the plots of land is the experimental unit. *(ii)* In the hospital, the patients may be the experimental unit *(iii)* A group of pigs in a pen may be the experimental unit in case of drug experiment.

Q.6. What is experimental material ?

Answer

Entire collection of experimental units is known as experimental material. For example the whole field is experimental material in case of an agricultural experiment.

Q.7. Explain clearly the term "experimental error".

Answer

A fundamental phenomonon in a replicated experiment is the variation in measurements made on different experimental units even when they get the same treatments. A part of this variation is systematic and can be explained where as the remainder is to be taken as to be of random nature. "The varaition in yield observation from plot to plot which is due to random (or chance or non assignable) factors and beyond human control is known as experimental error". It is a techincal term and it does not mean a mistake but includes all types of extraneous variation due to

(i) Inherent variability in the experimental material

(ii) Lack of uniformity in methodology of conducting the experiment.

(iii) Lack of representativeness of the sample to the population under consideration.

Q.8. What do you mean by the term treatment in the subject of design of experiment ?

Answer

Various objects (or procedures) to be compared in a comperative experiment, are termed as treatments. For example in field experimentation, the different fertilizers or different varieties of a crop or different methods of cultivation are treatements.

Q.9. What is the importance (use) of experimental error?

Answer

An experimental error provides the basis for confidence to be placed in the inference drawn from the experimental results about the population characteristcs. So it is important to estimate and control the experimental error.

Q.10 What are the basic principles of design of experiment through which an experimental error is controlled and estimated ?

Answer

An experimental error is controlled by the principle of local control and it can be estimated only by the replication.

Q.11. How the precision of an experiment is measured ?

Answer

Precision of an experiment $= \frac{1}{\sigma_{\bar{x}}^2} = \frac{n}{\sigma^2}$

where n : is the number of repetition of observation

σ^2 : is the variance of individual observation in the population

$\sigma_{\bar{x}}^2$: is the sampling variance of the sample estimator (Sample mean $\bar{x}$)

Q.12. How the precision of an experiment can be increased ?

Answer

The expression for the precision of an experiment given in question number 11 shows that the precision can be increased in the following two ways.

(i) By increasing the number of observation n i.e. by increasing the number of replication.

(ii) By reducing the variability σ^2 which can be done (controlled) by using the principle of local control.

Q.13. In what way an experiment should be conducted to give a valid conclusion.

Answer

To get a valid conclusion from the experiment, it is essential to control all the other variables except that or those under consideration i.e. the experiment should be conducted in an idential condition with respect to all the other variables except those under study.

Q.14. What do you mean by a contrast of observations ?

Answer

Let $y_1, y_2, \ldots\ldots, y_n$ denote *n* observations or any other quantities. The linear function $C=\sum_{i=1}^{n} li\ yi$

where l'6 are given constants such that $\sum_{i=1}^{n} li=0$, is called a contrast of yi's.

Q.15. Why the restriction $\sum_{i=1}^{n} li=0$ is imposed on a contrast ?

Answer

The restriction $\sum_{i=1}^{n} li=0$ makes the contrast C to be able to have comparision among the y values.

Q.16. Let $y_1, y_2, \ldots\ldots y_n$ are independent random variables with common mean μ and variance σ^2. Then what is the expected value and variance of the random variable C ?

Answer

$$E(c)=E(\sum_{i}^{n} li\,yi)=\sum_{i}^{n} li\,E(yi)=\mu\sum_{i}^{n} li=\mu x0=0, \quad (\therefore \sum_{i}^{n} li=0)$$

$$V(c)=\sum_{i}^{n} li^2\ V(yi)+2\sum_{l=i} li\,li\ cov(yi, yi)$$

$$=\sigma^2\sum_{i=1}^{n} li^2+0 \ (\therefore cov(yi\ yi)=0)$$

Q.17. What is the condition for orthogonality of two contrasts ?

Answer

Let there be two contrasts A and B defined as

$$A=\sum_{i=1}^{n} \lambda ixi, \quad \sum_{i=1}^{n} \lambda i=0 \quad \text{and}$$

n n

i i

i=1 i

Then the two contrasts A and B are said to be orthogonal if

$$\sum_{i=1}^{n} \lambda_i \mu i = 0$$

Q.18. Define an elementary contrast.

Answer

Let A be a contrast defined as

$$A = \sum_{i=1}^{n} \lambda i y i, \quad \sum_{i=1}^{n} \lambda i = 0$$

Then the contrast A is said to be an elementary contrast if λt's is have only two non zero elements as ±1.

Q.19. Write a set of three mutually orthogonal contrasts.

Answer

(i) $A = y_1 + y_2 - y_3 - y_4$

(ii) $B = y_1 - y_2 - y_3 + y_4$

(iii) $C = y_1 - y_2 + y_3 - y_4$

Remarks

(i) All the three functions A, B & C are contrasts.

(ii) These contrasts are mutually orthogonal

(iii) Each of the three contrasts A, B, C is an elementary contrast.

Q.20. What is uniformity trial ?

Answer

A trial conducted on a agricultural field in which the field (experimental material) is divided into small units (plots) and the same treatement is applied on each of the units and their yields are recorded, is known as uniformity trial.

Q.21. What is fertility contour map ?

Answer

In a uniformity trial, a graphic picture of the nature of the soil fertility variation is obtained by joining the points of equal fertility through line. This graphic picture is known as fertility contour map.

Remark : Uniformity trials also give us some idea about the shape and size of the plot to be used.

Q.22. What should be the shape of the plots and blocks and how they must be arranged with respect to fertility gradient in the field?

Answer

To get the maximum precision in the experiment, the plots should be rectangular in shape with their long side parallel to the direction of fertility gradient and the blocks should be arranged one after the other along the fertility gradient as shown in the fig.

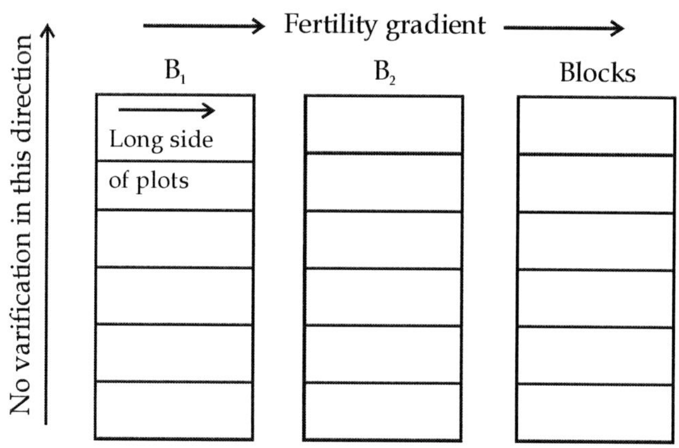

Q.23. What is the critical difference between the two terms accurancy and precision of the measurements and their estimates.

Answer

The accuracy of a measurement signifies the closeness with which a measurement approaches to the truth value. The lesser is the deviation from true value, the greater is the accuracy of an measurement. But the precision denotes the closeness with which a measurement approaches to the average value. The lesser is the variability of the measurement from the average value, the greater is the precision of these measurements and their estimates.

Q.24. What are the three basic principles of design of experiment ?

Answer

Professor R. A. Fisher pioneered the study of experimental designs with his classical boon "The Design of Expt". According to him, the basic principles of the design of experiments are

(i) Replication

(ii) Randomisation and

(iii) Local control

Q.25. What do you mean by replication ?

Answer

Replication means the repetition of the treatements under investigation.

Remarks

(i) Using replication, an experimentor is able to average out the influence of the chance factors on different experimental units.

(ii) The repetition of treatments results in more reliable estimate than that which is not possible with a single observation.

Q.26. What is the chief merit of replication ?

Answer

The replication serves to reduce the experimental error and also enable us to obtain more precise estimates of the treatment effects.

Q.27. What is the relation between the precision of an experiment and number of replications of a treatment ?

Answer

If a treatment is replicated *r* times, the sampling variance of its mean effect is $\frac{\sigma^2}{r}$ *where* σ^2, the variance of individual plot is estimated from error variance. Thus the precision of the experiment is given as

$$\text{precision} = \frac{1}{\text{variance}(\bar{x})} = \frac{r}{\sigma^2}$$

which shows that precision is directly in proportion to the number of replication.

Q.28. What is the role of replication in the estimation of experimental error ?

Answer

The most important role of replication is to provide an estimate of the experimental error without which we can not test the significance of the difference between any two treatments.

Q.29. How the estimate of the experimental error is obtained by considering the observations in different plots ?

Answer

The estimate of the experimental error is obtained by considering the differences of observations taken on plots receiving the same treatment in different replications and there is no other alternative of obtaining this estimate.

Q.30. A general rule is to get as many replications which will provide at least 12 degree of freedom for error. Why it is so ?

Answer

In any design the error degree of freedom must be at least 12 following the fact that the values of F statistic do not decrease repeatedly beyond √2 = 12, the error d. f. Usually one should not use less be than 4 replications

Q.31. What is an approximate idea to be used to find the minimum number of replications required to detect the given difference between two treatments at certain level of significance ?

Answer

The required approximate idea is obtained by applying normal test using z statistic defined as

$$Z=\frac{(\bar{x}_1-\bar{x}_2)}{\sigma\sqrt{2/r}}=\frac{d\sqrt{r}}{\sqrt{2.\sigma^2}}$$

However for small d.f., z is not normally distributed but it follows student's distribution. In that case we have

$$t=\frac{d\sqrt{r}}{\sqrt{2.s^2}}, \qquad \hat{\sigma}^2=s^2$$

$$r=\frac{2\,t_\alpha^2\,\sigma^2}{d^2}$$

which gives the approximate minimum number of replication required to detect a specified difference d between two treatments at α% level of significance.

where s^2 is the error variance per unit, α is value of student's t at α% level of significance corresponding to error d.f.

Q.32. What do you mean by randomisation?

Answer

The process of assigning the different treatments to the various experimental units in a purely random way is known as randomisation.

Q.33. What are the various uses of randomisation.

Answer

The principle of randomisation is essential for a valid estimate of experimental error and also to minimum the bias in the results obtained from an experiment.

Q.34. Using what principle or technique, the assumption of independence of errors in the model of ANOVA is fulfilled?

Answer

The assumption of independence of errors in the model of ANOVA is fulfilled by applying the technique of randomisation.

Remarks : The purpose of randomisation is to generate the random variables which are needed to perform a design of experiment.

Q.35. How can you assure that the error components involved in ANOVA model is purely random in nature ?

Answer

By using the principle of randomisation, one can assure for the random nature to be followed by the error components in ANOVA model.

Q.36. How one can assure that the average effect of sources of variation beyond human control in the experiment (error component) on any group of unit is zero.

Answer

One can assure that the average effect of error components on any group of unit or treatment will be zero only by applying the principle of randomisation.

Q.37. How the validity of the statistical tests of significance is maintained through randomisation ?

Answer

The statistical tests of significance (eg t test and F test for homogeneity of several means) depends on the fact that the statistic under consideration obeys some statistical distributions. Randomisation provides the logical basis for that and make it possible to draw rigorous inductive inferences. This assumption of randomisaton is necessary, since $S.E(\bar{x})=\frac{\sigma}{\sqrt{n}}$ for random sampling only.

Q.38. What do you mean by local control ?

Answer

The process of reduncing the experimental error by dividing the relatively heterogeneous experimental area into homogenous blocks, is known as local control.

Q.39. Does the local control increase the efficiency of a design? If yes then how ?

Answer

Yes. By using the local control, the experimental error is reduced and hence the efficiency of the design is automatically increased.

Q.40. What is the use of replication in collaboration with local control ?

Answer

Replication is used with local control to reduce the experimental error and hence the efficiency of the design is increased to a great extent.

Q.41. What is the joint importances (use) of replication and randomisation ?

Answer

By applying the replication and radomisation jointly, one can also find the estimate of experimental error in a valid way.

Q.42. What are the different ways to reduce the experimental error?

Answer

1. Using local control
2. Using confounding technique in factional experiment.

3. Use of one or more auxilary variable in analysis of covariance will also reduce the experimental error.
4. The proper choice of the size and shape of the experimental plots and blocks will also reduce the experimental error.

Chapter 2

ANOVA & Three Basic Design of Experiments

Analysis of Variance

Q.1. What is the definition of analysis of variance ?

Answer

According to Prof. R. A. Fisher, "ANOVA" is the separation of variance ascribable to one group of causes from the variance ascribable to another group of causes.

Q.2. What is the basic purpose of ANOVA technique ?

Answer

The basic purpose of ANOVA technique is to test the homogeneity (equality) of several means.

Q.3. Who did introduce first time, the term ANOVA in agricultural data ?

Answer

Prof. R. A. Fisher introduced first time the term ANOVA in agricultural data in the year 1990.

Remark : ANOVA technique is now frequently used in testing the linearity of fitted regression line or the significance of the correlation ratio η.

Q.4. What are the assumptions of ANOVA ?

Answer

The following assumptions should be followed by the observations to apply the ANOVA technique.

(i) The variate value y_{ij} should be random in nature otherwise their means μ_{ij} can not be estimated.

(ii) The variate values y_{ij} should be linear combination (additive) of the effects of several factors (e.g. treatments and error components) such as

$$y_{ij} = \mu + t_i + b_j + e_{ij}$$

(iii) The error e_{ij} should be randomally independently and normally distributed with zero mean and constant variance v $(e_{ij}) = \sigma_e^2$ so that we can get an unbiased estimate of σ_e^2 even when the hypothesis of homogeneity of means is discredited.

Q.5. Why the variate values y_{ij} should be homosedastic in ANOVA technique ?

Answer

The variates should be homosedastic otherwise one can not obtain independent estimate of common variance σ_e^2.

Q.6. Why the variates (parent population) should be normally distributed ?

Answer

The variates should normally be distributed because it is necessary to apply the F test even if the variates are random.

Q.7. What type of functional relationship is followed by degrees of freedom of the various factors.

Answer

The d.f. due to various factors (scores of variation) should be additive in nature i.e. total d.f. in ANOVA must be equal to the sum of d.f. due to various components of variation.

Q.8. What will be the value of F statistic if the null hypothesis Hσ is true ?

Answer

In case if the null hypothesis Hσ is true, theoretically the value of $F = S_t^2 / S_E^2$ is equal to 1 (unity) otherwise it will be greater than unity.

Remarks : (i) Under null hypothesis, by cochran theorem $\frac{S_t^2}{\sigma_e^2}$ and $\frac{S_E^2}{\sigma_e^2}$ *are independently distributed as* X^2 *with (t-1) and (n-t) d.f. respectively.*

(ii) The ratio $F = \frac{(s_t^2/\sigma_e^2)/(t-1)}{(s_E^2/\sigma_e^2)/(n-t)} = \frac{s_t^2/(t-1)}{s_E^2/(n-t)}$

(S_t^2 *&* S_E^2 *are due to treatment and error) follows F distribution with (t-1), (n-t) d.f.*

Q.9. Is it necessary for the components of variation (eg. treatment and error terms) to follow the normal distribution.

Answer

Yes, if, the treatments and error terms follow the normal distribution then only in that case $\frac{S_t^2}{\sigma_e^2}$ and $\frac{S_E^2}{\sigma_e^2}$ will follow χ^2 distribution which is necessary for applying F distribution *where* F is defined as the ratio of two independent mean. Chi squares.

Important Result : Although S_t^2 and S_E^2 add up to the total sum of square (S_{total}^2), the corresponding d.f. are also additive, their mean square (variance) do not add up to the total mean sum of square (M. S. S.). The technique though known as ANOVA is in fact the analysis of SS.

Q.10. What do you mean by a statistical model ?

Answer

A statistical model is actually a linear function of the effects of the different levels of a number of factors involved in an experiment along with one or more error terms.

Q.11. Write the definition of fixed effect mode, mixed effect model and random effect mode.

Answer

Fixed Effect Model : A model in which each of the factors has fixed effects and only error effect is random, is known as fixed effect model.

Mixed Effect Model : A model in which some factors have fixed effects and some have random effect, is known as mixed effect model.

Random Effect Model : A model in which all the factors have random effect, is known as random effect model.

Q.12. What is the main objective in the fixed effect model ?

Answer

The main objective in the fixed effect model is to estimate the effect of different factors considered in the study.

Q.13. What is the main objective in the random effect model ?

Answer

In a random effect model, the main objective is to find the estimate of the variability among the effects of different factors.

Q.14. What do you mean by one way classified data ?

Answer

When a set of observations are distributed over the different levels of only one factor, then such type of distribution (classification) of observation is known as one way classification. For example the observations put in the form of complete randomised design (CRD) is one way classified data.

Q.15. What do you mean by two way classified data ?

Answer

If we classify the observations with respect to two factors say A and B simultaneously at a number of levels t and r respectively than the classification is known as two way classification. For example the data put in the form of randomised block design (RBD) is said to be two way classified data.

Q.16. Write the model for one way classified data ?

Answer

The model is

$$y_{ij} = \mu + \beta_i + e_{ij} \quad i = 1, 2, \ldots\ldots\ldots, k$$
$$j = 1, 2, \ldots\ldots\ldots, ni$$

where μ : is the general mean

β_i : is the effect due to ith factor

$eij \sim N(o, \sigma_e^2)$

Q.17. What are the estimates of μ and β_i ?

Answer

μ and β_i are estimated as

$$\hat{\mu} = \frac{1}{n} \sum_{1}^{K} \sum_{j}^{ni} y_{ij} = \bar{y}..$$

$$\hat{\beta}_i = (\bar{y}_{i.} - \bar{y}_{..})$$

Q.18. What is the unbaised estimate of error variance σ_e^2 ?

Answer

Let S_E^2 be the error sum of square with (n-k) d.f. then error mean square $s_E^2 = \frac{S_E^2}{(n-k)}$ and $E(s_E^2) = \sigma_e^2$ which shows that error M. S. (s_E^2) is an unbaised estimate of σ_e^2.

Q.19. Let M.S.A. denotes the mean square due to factor (treatment) A. Then what is the expected value of M.S.A.?

Answer

We have

$$E\,(M.\,S.\,A.) = \sigma_e^2 + \phi\,(A_1, A_2, \ldots\ldots\ldots\ldots, A_k)$$

where $\phi\,(A_1, A_2, \ldots\ldots\ldots\ldots, A_k)$ is a variance like function of Aμ's. This function takes the value zero when Hσ is true otherwise it is a positive quantity.

Q.20. Under what condition, the treatment mean square (MSA) is an unbiased estimate of error variance ?

Answer

When null hypothesis H_0 is true, then treatments mean square an unbiased estimate of error variance.

Q.21. When the null hypothesis is rejected in case of ANOVA, then how one should proceed further ?

Answer

When the null hypothesis is rejected, than one should use the student's t test to test the pairwise equality of treatment means. For doing so one should calculate the critical difference (least significant difference).

Q.22. How the critical difference is calculated ?

Answer

Let r_1 and r_2 be the numbers of replication for two treatment t_1 and t_2 than we have the student's t statistic as

$$t\alpha/2 = \frac{(\bar{t}_1 - \bar{t}_2)}{\sqrt{\left(\frac{1}{r_1}+\frac{1}{r_2}\right)MSE_{error}}}$$

$$(t_1 - t_2)f = t_{\frac{\alpha}{2}}\sqrt{\left(\frac{1}{r_1}+\frac{1}{r_2}\right)MSE_{error}}$$

$$\text{or C.D} = t_{\frac{\alpha}{2}}\sqrt{\frac{2}{r}MSE_{error}} \quad \text{if } r_1 = r_2 = r$$

where $t_{\alpha/2}$ is the t value at $\alpha/2$ level of significance corresponding to error d.f. Thus if the difference between any two treatment means $(\bar{t}_i - \bar{t}_j)$ is greater than the critical difference then the $H_0 : (\bar{t}_i = \bar{t}_j)$ is rejected. at $100\alpha\%$ level otherwise it is accepted.

Remarks : $t_{\alpha/2}$ is the upper $\alpha/2$ point of the t distribution at error d.f.

Q.23. Can we apply the change of origin and scale to the original data in case of ANOVA technique ?

Answer

Yes, we can apply the change of origin or change of scale or change of origin and scale both to the original data in case of ANOVA technique to make the calculation easy and speedy.

Q.24. Why the conclusion obtained from the ANOVA based on transformed data (change of origin or scale or both) is same as based on original data ?

Answer

Since the variance is independant of change of origin and again the ratio of variance (F statistics) is independent of scale also and hence the final conclusion based on transformed data and original data will be the same.

Q.25. Write and explain the statistical mode for two way classified observations with one observation per cell.

Answer

We have the statistical model as

$y_{ij} = \mu + \alpha_i + \beta_i + r_{ij} + e_{ij}$

where μ : is the general mean effect present in all observations. α_i : is the effect due to i^{th} level of the factor A and common to all observations of ith level of the factor A

β_j : is the effect due to J^{th} level of the second factor B and common to all observations to the Jth level of B.

r_{ij} : is the interaction between i^{th} level of A and J^{th} level of B. *Result* : r_{ij} (the interaction) is not present in the ith level of A or in the Jth level of B if A and B are not taken together.

Q.26. If the joint effect of A_i and B_j is different from the sum of effects due to A_i and B_j taken individually then distinguish whether interaction is present or not ?

Answer

Under the situation described here, the interaction r^{ij} will be present.

Q.27. Write the unbiased estimate of μ, α_i and β_j.

Answer

μ, α_i and β_j will be estimated unbiasedly as

$$\hat{\mu} = \overline{y}$$

$$\hat{\alpha}_i = \overline{y}_i - \overline{y}$$

$$\hat{\beta}_i = \overline{y}_j - \overline{y}$$

Q.28. What will be the functional form of the expectation of MSA, MSB and MSE.

Answer

We have the following results

$$E(MSA) = \sigma_e^2 + \phi(\alpha_i^2)$$

$$E(MSB) = \sigma_e^2 + \gamma(\beta_j^2)$$

$$E(MSE) = \sigma_e^2$$

where $\phi(\alpha_i^2)$ and $\gamma(\beta_j^2)$ will be the terms likes variance

Q.29. When the null hypothesis for testing the significance of two factors A and B both are true then what will be the value of $\phi(\alpha_i^2)$ and $\lambda(\beta_j^2)$ xx ?

Answer

In case if the null hypothesis for both the factors A and B are true the value of $\phi(\alpha_i^2)$ and $\gamma(\beta_j^2)$ will be zero and

E (MSA) = σ_e^2

E (MSB) = σ_e^2

Completely Randomised Design (C.R.D.)

Q.30. Which one is the simplest design used. ?

Answer

The CRD is the simplest design out of all the existing designs.

Q.31. What are the two basic principles which are followed by CRD?

Answer

The two basic principles to be followed by CRD are randomisation and replication.

Q.32. What is the condition which must be followed by the experimental material to apply the CRD ?

Answer

To apply the CRD, the experimental material should be homogeneous.

Q.33. In which design, the complete fexibility regarding the number of treatments and number of replication is allowed ?

Answer

Such type of fexibility is allowed in CRD.

Q.34. Which one of the three basic principles assures that the extrameous factors do not continuously influence one treatment ?

Answer

Randomisation makes this surity.

Q.35. In which design one can make the maximum use of the experimental units ?

Answer

C.R.D.

Remark : C.R.D. provides the maximum number of d.f. for the estimation of error variance which increase the sensitivity or the precision of the experiment for small experiment i.e. for experiment with small number of treatment.

Q.36. Suppose one has to conduct an experiment in laboratory, and in some methodological studies eg. in physics, chemistry, cookery, in chemical and biological experiment in some green house studies. Which design will be suitable ?

Answer

C.R.D.

Q.37. Let the two treatments (i, j) be replicated (r_i, r_j) times in C. R.D. What will be the standard error of the difference between (i, j) the treatment means ?

Answer

S. E. $(\bar{x}_i - \bar{x}_j) = \sqrt{s_E^2\left(\frac{1}{r_i} + \frac{1}{r_j}\right)}$

where s_E^2 is the estimate of error variance σ_e^2.

Randomised Block Design (R. B. D.)

Q.38. What is randomised block design ?

Answer

When the experimental material is not entirely homogeneous, we divide the whole material into homogeneous group of experimental units by adopting the principle of local control and randomisation is done separately within each group (block). Such a layout is known as randomised block design.

Remarks:

- The R.B.D. uses all the three basic principles enumerated by Fisher R.A.
- In R.B.D. each treatment is replicated equal number of times.
- Analysis of R.B.D. is same as that of two way classified data with one observation per cell.
- At all the stages of the experiment in R.B.D., the object is to keep the experimental errors within each group as small as possible.

Q.39. Let the observation on treatment 1 in block 1 is missing in R. B. D. Then what will be the substitute of the missing value to analyse the data in case of R.B.D.

Answer

Let x be the substitute for the missing data then we have

$$x = \frac{(tT^1 + rB^1 - G^1)}{(r-1)(t-1)}$$

where r : is the number of replications or blocks

t : is the number of treatments or block size.

T^1 : is total of treatment containing the missing value

B^1 : Total of block containing the missing value

G^1 : Grand total of all the available observations

Q.40. What will be the error and total d.f. if one observation is missing in R.B.D.

Answer

Error d.f. = (t-1) (r-1) - 1

Total d.f. = (rt-1) - 1

Q.41. What will be the efficiency of R. B. D. over C. R. D. ?

Answer

For this, first we calculated error MSE (s_E^2) of C. R. D. i.e. if R. B. D. were conducted as C RD i.e. if blocks were not formed, then

$$s_e^2(\text{C.R.D}) = \frac{n_r S_r^2 + (n_t + n_e) S_e^2}{(n_r + n_t + n_e)}$$

where n_r, n_t, n_e are the d.f. for replication, treatment and error respectively and let s_E^2 be error MSS of RBD.

Then efficiency (RBD over CRD) = $\dfrac{S_e^2(\text{CRD})}{S_e^2(\text{RBD})} \times 100$

Q.42. If more than two observations are missing in RBD then what technique will be more appropriate to estimate then ?

Answer

In the situation described above, the Bartlett's missing plot technique will be more appropriate.

Q.43. What will be the error d.f. in case of two missing observation in RBD ?

Answer

Error d.f. = (r-1) (t-1) - 2

Remark : Total MSS and error MSS always gives unbiased estimate of σ_e^2, the error variance

Q.44. Under what situation Row MSS and Column MSS are unbiased estimate of σ_e^2 ?

Answer

The above result hold when the null hypothese for row & column are true.

Remark : Under the assumption of normality of parent population we get by cochran's theorem that

$\frac{S^2_{ROW}}{\sigma^2_e}, \frac{S^2_{COL}}{\sigma^2_e}$ and $\frac{S^2_{Error}}{\sigma^2_e}$ are mutually and independently distributed as χ^2 with respective d.f.

Q.45. Let F be defined as the ratio $F = \frac{S^2_t}{S^2_e}$ than give the name of distribution followed by F

Answer

The ratio F will follow central F distribution with (t - 1), (t-1) (r-1) df.

Q.46. What will be the formula to calculate the C.D. between two treatments means in RBD.

Answer

C. D. (between two treatment means)

$$= t_\alpha (r-1)(t-1) \times \sqrt{\frac{2S^2_e}{r}}$$

Q.47. Suppose some extra replications is required for two treatments t_1 and t_2 in RBD. They are replicated r_1 and r_2 times respectively. Then what will be the S.E. for the difference of $(\bar{t}_1 - \bar{t}_2)$

Answer

S. E. $(\bar{t}_1 - \bar{t}_2) = Se\sqrt{\frac{1}{r_1} + \frac{1}{r_2}}$ *where* S^2_e is the error variance

Latin Square Design

Q.48. What kind of layout is known as L.S.D. ?

Answer

In agricultural field experiment where fertility contour is not known, the grouping of experimental unit is done in such a way that it controls the variation simultaneously in two directions at right angles. Such a layout is known as LSD.

Q.49. Where L.S.D. is used ?

Answer

L.S.D. has been successfully utilised in industry, lab experiment and Agrl. field experiment.

Q.50. How the arrangement of latin square design is done ?

Answer

In L.S.D., the number of treatments is equal to the number of replications. Thus in case of m treatments there has to m x m = m^2 experimiential units. The m^2 experimental units will be arranged in a square.

The m treatments are then allocated at random to these rows and columns such that every treatment occurs once and only once in each row and each column. Such a lay out is known as mxm latin square.

Q.51. Write an example of 4 X 4 LSD taking A, B, C & D as four treatments ?

Answer

	A	B	C	D
Example	B	C	D	A
	C	D	A	B
	D	A	B	C

Remark : LSD utilizes all the three basic principles of design of experiment named randomisation, replication and local control.

Q.52. L.S.D. is an incomplete three way layout. In what way it is advantages over the complete three way layout ?

Answer

An incomplete three way layout utilizes only m^2 experimental units instead of m^3 experimental units utilized in case of complete three way layout.

Q.53. What is the approximate number of treatments generally used in L.S.D. ?

Answer

L.S.D. is suitable for the number of treatments between 5 to 10.

Q.54. What will be the error d.f. in case of m X m L.S.D. ?

Answer

error d.f. = (m-1) (m-2)

Q.55. Write the linear model for L.S.D.

Answer

The linear model is

$Y_{ijk} = \mu + \alpha_i + \beta_j + t_k + e_{ijk}$ (i, j, k = 1, 2, ..., m)

where α_i, β_j & t_k are the effect of ith row, jth column and K^{th} treatment, e_{ijk} is error term N (o, σ_e^2).

Chapter 3

Incomplete Block Design (IBD)

Q.1. What is an incomplete block design?

Answer

An incomplete block design is that design in which each block does not contain a complete set of treatments. So an incomplete block design is one in which the block size is less than the number of treatments so that all the treatments can not be accommodated in the same block.

Q.2. What is the definition of an incomplete block design ?

Answer

An incomplete block design is one having v treatment, and b blocks each of size $K < V$, such that each of the treatment is replicated r times and each pair of treatments occurs once and only once in the same block. v, b, r and k are known as the parameters of the IBD.

Q.3. What is balanced incomplete block design ?

Answer

An arrangement of v treatments in b blocks each of size $K < V$ is known as BIBD if

(i) Each block contains K distinct treatments.

(ii) Each treatment occurs in r blocks

(iii) Each pair of treatments occurs together in λ blocks.

Q.4. What are the parameters of BIBD ?

Answer

The integers v, b, r, k, and λ are known as the parameters of BIBD.

Q.5. What is the meaning of the parameter λ in BIBD ?

Answer

λ denotes the number of blocks in which two treatments occurs together or number of times any two treatments occurs together in the design.

Q.6. What is the necessary condition for the existence of BIBD ?

Answer

The following parametric relations serve as the necessary condition for the existence of BIBD

(i) $vr = bk$

(ii) $\lambda\,(v-1) = r\,(k-1)$

(iii) $b \geq v$ (Fisher's inequality)

Q.7. What is an incidence matrix related to BIBD ?

Answer

Given a BIBD there exists always an incidence matrix N_{vxb} as

$N_{vxb} = (n_{ij})$ such that

$n_{ij} = 1$ if i^{th} treatment occurs in j^{th} block

0 otherwise

N_{vxb} will be of the form

		block				
		1	2	....	b	Total
	1	n_{11}	n_{12}	....	n_{1b}	r
	2					r
	.					.
Treatment	.					.
	i	ni1	ni2	....	nir	γ
	.					.
	.					.
	v	n_{vx1}	n_{vx1}	....	n_{vxb}	r
	Total	K	K	....	K	

Q.8. What will be the value of inner product of any two rows of an incidence matrix ?

Answer

The inner product of any two rows of N is equal to λ. Which is clear from the following example. Let the design be

$v = b = 4, r = k = 3, \lambda = 2$

$$N = \begin{bmatrix} 1 & 1 & 1 & 0 \\ 1 & 1 & 0 & 1 \\ 1 & 0 & 1 & 1 \\ 0 & 1 & 1 & 1 \end{bmatrix}$$

Inner product of any (1 & 2 no)

two rows = 1X1 + 1X1 +1X0 + 0X1

= 1 + 1

= 2 = λ

Remark : The parametric relations are only the necessary conditions and not the sufficient in the sense that even though all the parametric relations hold true, it may not be possible to arrange v treatments in b blocks in the above mentioned fashion.

Q.9. If r = λ for any BIBD then BIBD will reduces to what kind of design ?

Answer

If for any BIBD, λ = r then BIBD reduces to a complete block because in that case the block size will be equal to the number of treatment.

Q.10. What is resolvable BIBD ?

Answer

A BIBD is said to be resolvable if the *b* blocks can be divided into *r* groups each containing $\left(\frac{b}{r}\right)$ blocks such that each group contains every treatment precisely once $\left(\frac{b}{r}\right)$ is an integer. It means that each group is one replicate.

Q.11. When a resolvable design is said to be afine ?

Answer

A resolvable design is said to be affine resolvable if any two blocks from different groups have $\left(\frac{k^2}{v}\right)$ treatments common. Here $\left(\frac{k^2}{v}\right)$ is an integer and b= v +r-1 is a necessary condition for a resolvable design to be affine resolvable.

Q.12. What is symmetrical BIBD ?

Answer

A BIBD is said to be symmetrical if v = b and r = k so there are only independent parameters. In other words, a design is said to be symmetrical if any two blocks interset at a constant number of treatments and that constant is λ.

Q.13. Define a complementary design.

Answer

Let there be a BIBD with parameter D (v, b, r, k, λ). Then there exists another BIBD with parameters $\overline{D}$ (v^1, b^1, r^1, k^1, λ^1) such that

$$\left.\begin{aligned} V^1 &= v \\ b^1 &= b \\ r^1 &= b-r \\ k^1 &= v-k \\ \lambda^1 &= b - 2r + \lambda \end{aligned}\right\}$$

and $\bar{D}$ will be called as complementary design and it will also be a BIBD.

Remark : If B_1, B_2, ..., B_b are the blocks of a BIBD (say D) then $\bar{D}$ (complementary design of D) will consist of blocks $\bar{B}_1, \bar{B}_2, \ldots\ldots\ldots, \bar{B}_b$ where $\bar{B}_i$ is the complementary block of the block B_i in D.

Q.14. What is residual design ?

Answer

Let D be a symmetrical BIBD consisting of treatments (block) S_1, S_2, ... S_v and with the parameters v = b, r = K & λ. Let us select one of these blocks say S_1, and delete from each of the S_2, S_3, ... S_v. blocks, the λ treatment which are common with S1. The deleted set $S_2^*, S_3^*, \ldots\ldots\ldots\ldots S_v^*$ form a design which is called residual design D* in which parameter v^1, b^1, r^1, k^1, λ^1 are given by

$v^1 = v\text{-}k^1$

$b^1 = v\text{-}1 = b\text{-}1$

$r^1 = k^1$

$k^1 = K\text{-}\lambda$ and $\lambda^1 = \lambda$

and derived design D* is also a BIBD

Q.15. What is derived design ?

Answer

We may also form another design D′ by taking other sets S_2' S3′ .... Sv′ with the λ symbols that S_2, S_3, ..., S_v respectively

have in common with S_1. This new formed design is known as derived design.
This design is also a BBD with parameters
$v_1 = k, \quad b_1 = v-1, \quad \lambda_1 = \lambda-1$

Q.16. When a design is said to be connected ?

Answer

A design is said to be connected if all the elementary contrast of treatments are estimable otherwise it is called disconnected.

Q.17. When a design is said to be balanced ?

Answer

A design is said to be balanced, if all the elementary contrast of treatments are estimated with the same precision.

Remarks :

(1) Latin square and RBD are balanced, connected and orthogonal designs.

(2) Efficiency factor of BIBD is less than the unity.

Partialy Balance Incomplete Block Design (PBIBD)

Definition : An incomplete block design in which all the elementary contrast are not estimated with the same precision, is known as partially balanced incomplete block design. The number v, b, r, K, λi (i = 1, 2, ..., m) are called the parameter of PBIBD.

Interblock and Intrablock Information

Treatment effects are affected by two type of variations.

(i) Variation due to random source operating within blocks are called intrablock variations.

(ii) Variation due to different fertility levels among subblocks within each replication is called interblock variation.

Chapter 4

Factorial Experiment

Preliminaries

It is an experiment, it is not a design. In an factorial experiment, the all possible combinations of two or more than two factors are considered. These different combinations of factors are treated as different treatments. After making the treatment combinations the experiment may be conducted in any one of desings say, CRD, RBD, LSD etc.

Q.1. What is factorial experiment ?

Answer

An experiment in which the effects of two or more than two factors, say variety, manure, data of sowing etc are considered together is called factorial experiment.

Q.2. What is a simple experiment ?

Answer

An experiment in which the effect of only one factor say variety or manure is considered at a time, is called simple experiment.

Q.3. What is pxq factorial experiment ?

Answer

A factorial experiment in which if p different levels of one factor say A and q different levels of other factor say B are considered together, is called a pxq factorial experiment.

Q.4. What do you mean by 2^n factorial experiment ?

Answer

By 2^n factorial experiment we mean that the effect of n factors (treatments) each at two levels has been considered in the experiment. *where* $n \geq 2$.

Q.5. Explain the main effect and interations taking two factors A and B each at two levels.

Answer

Let the two levels of two factors A and B be denoted by 0 and 1. Denote by small letters a, b and ab the treatment combinations and by (a), (b), and (ab) the mean treatment yields for treatment a, b and ab. Consider the following table.

Factor A →

Factor B ↓

	1st level a_0	**2nd level** a_1	**Response to A or Simple effect of A**
1st level b_0	a_0b_0 (1)	a_1b_0 (a)	(a) - (1)
2nd level b_1	a_0b_1 (b)	a_1b_1 (ab)	(ab) - (b)
Response to B or simple effect of B	(b) - (1)	(ab) - (a)	

Explanation for main effect of A

Simple Effect A at given b_0 level of B = $a_1b_0 - a_0b_0$

Simple Effect of A at given b_1 level of B = $a_1b_1 - a_0b_1$

Effect of A averaged over two levels of B

$= \frac{1}{2} [(a_1b_0 - a_0b_0) + (a_1b_1 - a_0b_1)]$

$= \frac{1}{2} [(a_1 - a_0) + (b_1 + b_0)]$

Thus main effect of A = $\frac{1}{2} (a_1 - a_0) (b_1 + b_0)$ and is denoted by A. Similarly B = $\frac{1}{2} (b_1 - b_0) (a_1 + a_0)$

Interaction AB : consider the difference of two simple effects of A.

$d = (a_1b_1 - a_0b_1) - (a_1b_0 - a_0b_0)$

$d = (a_1 - a_0) (b_1 - b_0)$

In case of two factors A and B being independent the differenced of two simple effect of A will be of order zero. If not, this will give an estimate of interdependence of two factors and it is known as interaction between A and B. Yates defined the interaction between A and B as

$AB = \frac{1}{2} (a_1 - a_0) (b_1 - b_0)$

Q.6. Can we call the simple effects, main effects and interactions as contrasts (or comparisons) ?

Answer

Really all the different type of effects are contrasts or comparisions.

Q.7. Are the above different type of contrasts othogonal to each other ?

Answer

Yes, it is easy to see that the contrast A, B, and AB are othogonal contrasts.

Remarks : In general if there are n factors A, B, C, D, E..... etc each at two levels 0 and 1 then the main effects and interaction may be represented by

$$X = \frac{1}{2^{n-1}} (a_1 \pm a_0)(b_1 \pm b_0)(c_1 \pm c_0)(d_1 \pm d_0)............(.......)$$

where the sign in each bracket is positive if the corresponding capital letter is not present in X and negative if it is present in X and the whole expression on the RHS is to be expanded algebraically and the yields are to be substituted in place of the corresponding treatment combination.

Q.8. Let the main effects of some factors be estimated from a factorial experiment as well as they be estimated from simple experiment as if the whole experiment is devoted only for that effect then which experiment will give more precise estimate of main effects ?

Answer

In case of estimation of main effects only, then both the factorial experiment as well as the simple experiment give equally efficient (precise) estimate of main effects.

Q.9. Let the average effects of A over various levels of B be found to be significantly different. What would be inferred about the interaction of two factors A and B ?

Answer

In this situation there will be interdependency as well as the presence of interaction AB between two factors A and B.

Results on 2^2 Factorial Experiment

Let there be two factors A and B each at two levels. We use the following notations.

(i) Treatment's mean yields are denoted as
(1), (a), (b), (ab)

(ii) Treatment totals are dented by
[1], [a], [b], [ab]

(iii) Factorial effects totals are gives as
[A] = [ab] + [a] - [b] - [1]
[B] = [ab] - [a] + [b] - [1]
[C] = [ab] - [a] - [b] + [1]

Q.10. If there are r replications in 2^2 experiment then how the SS due to main effect of A, B and interaction AB are calculated ?

Answer

We use the following formula to calculate them

SS due to main effect of A = $[A]^2/4r$

SS due to main effect of B = $[B]^2/4r$

SS due to interaction AB = $[AB]^2/4r$

and each will be with 1. df

Q.11. What will be the total and error d.f. in the above experiment?

Answer

Total d.f. = (4r - 1)

Error d.f. = 3 (r-1)

Q.12. If F_A, F_B and F_{AB} denote the variance ratio statistics then what will be their distribution ?

Answer

Each of the statistics F_A, F_B and F_{AB} follows central F distribution with (1, 3(r-1)) d.f.

Q.13. What would be the expressions for factorial effects (main and interaction effect) in terms of treatment total in 2^2 experiment with r replications.

Answer

Main effect of A = [A]/2r

Main effect of B = [B]/2r

Interaction AB = [AB]/2r

where [A], [B] and [AB] denote the factorial effect totals.

Q.14. Write the expressions for S.E. of factorial effect total and factorial effect means in 2^2 experiment with r replication.

Answer

S. E. ([A], or [B] or [AB]) = $\sqrt{4r\sigma_e^2}$

S. E. (Factorial effect mean) = $SE\left[\frac{\text{Factorical effect total}}{2r}\right]$

$$= \sqrt{\frac{4r\sigma_e^2}{4r^2}}$$

$$= \sqrt{\frac{\sigma_e^2}{r}}$$

where σ_e^2 is the error variance which is estimated by error mean SS viz. s_E^2

Q.15. What are the significant values for the factorial effect totals and factorial effect means at α% level of significance in 2^2 experiment with r replication ?

Answer

Let d_1 and d_2 be the corresponding significant values for factorial effect totals and factorial effect means.

$$d_1 = t_{\alpha\% \text{ (error d.f.)}} \sqrt{4rs_e^2}$$

$$= t_{\alpha\% \; 3(r-1)} \sqrt{4rs_e^2}$$

and $d_2 = t_{\alpha\%,\, 3(r-1)} \sqrt{\frac{S_e^2}{r}}$

Q.16. Why a special computational rule developed by F Yates is adopted to compute the factorial effect totals ?

Answer

The above method is adopted only to avoid the specific algebraic formulae in computation of factorial effect total.

Remark : We may interpret the second order interaction ABC as the difference in the interaction AB calculated at each of the two levels of C.

2^3 Experiment in r Replication

Let A, B & C be the three factors then there will be $2^3 = 8$ treatment combinations. They are 1, a, b, ab, c, ac, bc, abc.

Q.17. Let M denotes the mean yield of all the eight treatment combinations then what will be the formula to compute M.

Answer

$M = \frac{1}{8} [(a+1)\ (b+1)\ (c+1)]$

Q.18. What is the number of mutually orthogonal contrast in 2^3 factorial experiment?

Answer

In 2^3 factorial experiment, there are seven mutually orthogonal contrasts.

Q.19. What will be the error d.f. in 2^3 experiment with r replication.

Answer

Error d.f. = 7 (r-1)

Q.20. What is the S. E. of any factorial effect totals in 2^3 experiment

Answer

S. E. (Any factorial effect total) = $\sqrt{8r\sigma_e^2} = \sqrt{2^3\, r\, \sigma_e^2}$

Q.21. What is the S. E. of any factorial effect mean in 2^3 experiment.

Answer

S. E. (Any factorial effect mean) = $\sqrt{\frac{\sigma_e^2}{2r}} = \sqrt{\frac{2^3\, r\, \sigma_e^2}{(4r)^2}}$

Q.22. What will be the expression for interaction effect ABC in term of factorial effect total in 2^3 experiment?

Answer

Interaction ABC = [ABC]/4r

Q.23. In case of 2^n factorial experiment with r replication, what will be the general expression for SS due to each factorial effect ?

Answer

SS due to each factorial effect $= [\ ]^2/2^n r$

Where [ ] dentoes the factorial effect total

Q.24. What will be error d.f. in 2^n experiment with r replication

Answer

Error d.f. = $(2^n\text{-}1)\ (r\text{-}1)$

Confounding

Q.25. What is confounding ?

Answer

It is a device of reducing the block size by making unimportant one or more interaction contrasts identical with block contrast. The design for confounded factorial experiment may be called as incomplete randomised block design.

Q.26. How the confounding increases the precision of unfounded treatment combination ?

Answer

By the use of confounding technique the reduction of block size becomes effective in reducing the error variance and hence the precision of uncomfounded treatment combination is increased accordingly.

Complete Confounding : If the same interaction is confounded in all the replications, the confounding is said to be complete confounding.

Q.27. What is the advantage of sacrificing the information on confounded interaction ?

Answer

By sacrificing the information on confounded interaction the unconfounded (important) treatment combination are estimated with greater precision.

Q.28. What is the disadvantage of complete confounding ?

Answer

We can not get any information an complete confounded treatment combination.

Q.29. What is partial confounding ?

Answer

If the different interactions are confounded in different replications, then the confounding is said to be partial confounding.

Q.30. What is the advantage of partial confounding over complete confounding ?

Answer

In case of partial confounding, the interaction which is confounded in one replication can be estimated from the other replication and vice versa but in complete confounding the confounded interaction can not be estimated at all.

Q.31. What is assymetrical factorial experiment ?

Answer

If the number of levels of different factors (treatment combinations) considered in a factorial experiment is not same then it is said to be assymetrical or mixed factorial experiment.

Q.32. What is advantage of mixed factorial experiment ?

Answer

The chief advantage of the mixed factorial experiment is that they are more flexible in the sense that the level of factor can be chosen according to needs and conveniance of the experiment.

Q.33. What is the demerit of mixed factorial experiment ?

Answer

The construction of design and the method of analysis of such experiments are some what complicated.

Q.34. How the confounding technique is performed ?

Answer

The device of confounding consists in sub-dividing the replicate into two or more groups of equal size following certain rules by which we sacrifice some information on certain higher order interactions and allocating the treatments combinations of any group to any block at random.

Q.35. Let in case of complete confounding in 2^3 factorial experiment with r replication, each replicate is divided into two groups of equal size. What would be the d.f. for blocks as well as error.

Answer

Block d.f. = (2r-1)

Errors d.f. = 6 (r-1)

Treatment d.f. = $(2^3-1) - 1 = 6$

Remark : Effects totals of partially confounded effects are also termed as Interblock comparisions and those of unfounded effects as Intrablock comparision.

Q.36. In case of partial confounding in 2^3 factorial experiment with r replications each is divided into two groups of equal size. What would be the break up of d.f. in ANOVA.

Answer

Source	d.f.
Block	(2r-1)
Treatment	$(2^3-1) = 7$
Error	6 (r-1) - 1
Total	$(2^3\,r-1) = 8r - 1$

Q.37. What is a generalized interaction in factorial experiment?

Answer

Let there be 2^n factorial experiment with block size 2^k. The number of independent interactions confounded is (n -k). Then all possible products of (n - k) independent interactions confounded give us some more interactions which are known as generalized interactions.

Chapter 5

Split Plot Experiment

Q.1. What is split plot experiment ?

Answer

If one has to study two type of factors say A and B in the same experiment where A requires larger plot size and B requires smaller plot size then the structure of the experiment will be that of plots within blocks are known as split plots and blocks within replication are known as main (whole) plots. Such a layout of the experiment is known as split plot experiment.

Q.2. What is the randomisation producer used in split plot experiment ?

Answer

In split experiment, the main plot treatments are alloted at random to the main plots and the split plot treatments are alloted at random to the split plots.

Q.3. What is the difference in randomisation procedure between split plot experiment and ordinary two factors experiment in RBD ?

Answer

In split plot experiment randomisation is done separatory for the whole plot treatments and for the split plot treatments while in ordinary two factors experiment, all the treatment combinations of two factors are alloted at random to the plots of a block.

Remark : In split plot experiment, we utilize the interblock information from the start.

Q.4. What is the difference between split plot experiment and ordinary two factors experiment in RBD with efficiency point of view ?

Answer

In split plot experiment, the interaction of the whole plot treatments and split plot treatments are estimated more precisely (efficiently) than the main effects of the whole plot treatment. On the other hand the main effects and interactions are all tested equally efficient in the two factor experiment in RBD.

Q.5. Write some examples of main plot treatments and sub-plot treatments.

Answer

Main plot treatment : Sowing dates, different methods of ploughing, different method of irrigation.

Split (Sub) Plot treatment : Different varities, different manures, different dozes of a manure.

Q.6. How one can explain the similarity of split plot experiment with a confounded factorial experiment ?

Answer

If sub-plots are considered as plots and whole plots as block, we find that the difference among the whole plots are the same as the difference among the levels of the whole plot treatments. Hence one can say that the split plot experiment has confounded the main effects of the whole plot treatment.

Q.7. Write the model used in split plot experiment.

Answer

Let there be a main factor A at *p* levels, and sub factors B at *q* levels with *r* replication in a split plot experiment. Then the following model is used.

$$Y_{ijk} = \mu + r_i + a_j + \eta_{ij} + b_k + (ab)_{jk} + e_{ijk}$$

where i = 1, 2, ..., r

j = 1, 2, ..., p

k = 1, 2, ..., q

a_j, b_k and $(ab)_{jk}$ are fixed effects

n_{ij} = whole plot error

e_{ijk} = Subplot error

μ = General mean

r_i = Effect due to i^{th} replicated

Q.8. What will be the distribution of two errors η_{ij} and e_{ijk} ?

Answer

The $\eta_{ij} \sim N(0, \sigma^2_\omega)$ and eijk $\sim N(0, \sigma^2_s)$ independently.

Q.9. Why the joint effect of split plot treatment and replication has not been considered in the model ?

Answer

The joint effect (interaction) of split plot treatment and replication is assumed to be zero so this term is absent in the model.

Q.10. What is the structure of ANOVA.

Answer

Source of variation	d.f.	M.s.	F ratio
Replication	(r-1)		
Main plot treat A	(p-1)	MSA	MSA/Ea
Main plot error Ea	(r-1) (p-1)	MSEa = Ea	
Split plot treatment B	(q-1)	MSB	MSB/Eb
Interaction AB	(p-1) (q-1)	MS(AB)	MS(AB)/Eb
Split plot error Eb	p (r-1) (q-1)	MS Eb = Eb	
	r pq-1		

Q.11. What is the null hypothesis and test statistics to be used to test the significance of whole plot treatment ?

Answer

$H_0 : a_j = 0, (j = 1, 2, ..., p)$ i.e., ther is no significant difference between the whole plot treatments averaged over all the split plot treatments.

Test statistics is

$$F = \frac{MSA}{Ea} \sim F_{(p-1),(r-1)(p-1)}$$

Q.12. Write the null hypothesis and test statistics used to test the significance of split plot treatment ?

Answer

$H_0 : b_k = 0 \quad (K = 1, 2, ..., q)$ i.e. there is no signigicant difference between split plot treatments averaged over all the whole plot treatment

Test statistic used is :

$$F = MSB/Eb \sim F_{(q-1),\ p(r-1)(q-1)}$$

Q.13. Write the null hypothesis to test the interaction of whole plot treatments and split plot treatment and test statistic used.

Answer

$H_0 : (ab)_{jk} = 0$ i.e. that interactions of whole plot treatments and split plot treatments are all zero.

Test statistic used is

$$F = MS\ (AB)\ /\ Eb \sim F_{(p-1)\ (q-1),\ p\ (r-1)\ (q-1)}$$

Q.14. Out of the two errors Ea and Eb, which one is smaller than the other ?

Answer

Out of two errors, the split plot error Eb is smaller than the main plot error Ea.

Q.15. Usually the sub plot factor B and the interaction factor AB are estimated and tested more precisely than the main factor A. Why it is so ?

Answer

The above result holds good because the split plot error Eb is usually smaller than the whole plot error Ea.

Remark : It is possible to introduce second factor B requiring smaller experimental material along with A at little extra cost in split plot arrangement.

Q.16. The factor B is more important than the factor A. Then how one should allot the factors A and B to the whole plot and split plot in a split plot experiment ?

Answer

The more important factor must be alloted to the split plots in the experiment.

Q.17. Some times the error Ea may be two large. How it will affect the estimation and testing the significance of whole plot treatment ?

Answer

In that situation the estimation and testing of the whole plot treatment will be less precise.

Q.18. What is the estimates of standard error between two A means and two B means.

Answer

(i) SE (Difference between two A means e.g. $(a_j - a_j')$)

$$= \sqrt{\frac{2E_a}{rq}}$$

(ii) SE (Difference between two B means e.g. $(b_k - b_k')$)

$$= \sqrt{\frac{2E_b}{rp}}$$

Q.19. Write the estimate of S. E. of difference between two B means at the same level of A. e.g. $(a_j b_k - a_j b_k')$.

Answer

SE (Difference between two B means at the same level of A)

$$= \sqrt{\frac{2E_b}{r}}$$

Q.20 Write the estimate of S.E. between two A means at the same level of B or different level of B e.g. $(a_1\, b_1 - a_0\, b_1)$ or $(a_1\, b_1 - a_0\, b_0)$.

Answer

SE (Difference between two A means at the same level of B or different level of B)

$$= \sqrt{2\left[\frac{(q-1)E_b + E_a}{rq}\right]}$$

Remark : The experimental error for sub-plot treatment and interaction is smaller than that of the average experimental error over all treatment comparisons but it is same for split plot design and the corresponding factorial experiment in RBD.

Q.21. If the spliting of the sub-plot is further continued to the second order sub plots then the ultimate stage sub plot effects would be estimated and tested with the greatest precision. Why it is so ?

Answer

The ultimate stage sub plot effects would be estimated and tested with greatest precision because at that stage the local control would be most effective and the ultimate stage sub-plot would be compared under more homogenous condition than the other order sub-plots and the main plots.

Chapter **6**

Strip Plot Design/Split Block Design

Q.1. Define strip plot design ?

Answer

There is a variant of the split plot design in which both factors A and B will be applied to the larger strips by dividing the experimental field into as many rows as the number of levels of one factor A and into as many columns as the number of levels of second factor B. Then the two factors will be applied at random to the rows and columns. Such a layout is known as strip plot design.

Q.2. In some cases the treatments A and B may be of relatively little interest compared to the interaction AB. What design will be suitable in that case.

Answer

If both the treatments A and B require larger plots and the interaction AB is more important than the main effect A and B then strip plot design is suitable.

Q.3. Give some examples of treatments to be applied in strip plot design.

Answer

(1) Fertilizer and tillage treatments with ordinary farm implements

(2) Sparying treatments and tillage treatment with ordinary farm implements.

Q.4. What are the other appropriate terms to be used to denote strip plot design.

Answer

The other suitable terms to denote strip plot design may be split blocks, two way whole plots, strip trials or sub-units in strips etc.

Q.5. Write and explain the statistical model used in strip plot design.

Answer

The statistical model used in strip plot design is

$$Y_{ijk} = \mu + r_i + a_j + (ra)_{ij} + b_k + (rb)_{ik} + (ab)_{jk} + e_{ijk}$$

where r_i = i^{th} replicate or block design

a_j = effect of Jth level of factor A

b_k = effect of the Kth level of factor B

$$\sum_i ri = \sum_j aj = \sum_k b_k = 0 \quad \text{and} \quad \sum_{ij}(ra)_{ij} = \sum_{ik}(rb)_{ik} = \sum_{j,k}(ab)_{jk} = 0$$

e_{ijk} : error term.

Q.6. How many error terms are considered in the model ?

Answer

There are three error terms considered in strip plot design's model. They are as

(1) Error (a) obtained as Ea to be used to test the significance of main effect A

(2) Error (b) obtained as Eb to be used to test the significance of main effect B

(3) Error (c) obtained as Ec to be used to test the significance of the interaction AB effect.

Q.7. Write the structure of ANOVA Table.

Answer

Source of variation	d.f.	M.s.	F ratio
Replication	r-1		
A	(p-1)	MSA	MSA/Ea
Error (a) : Ea	(r-1) (p-1)	Ea	
B	(q-1)	MSB	MSB/Eb
Error (b) : Eb	(r-1) (q-1)	Eb	
AB	(p-1) (q-1)	MS(AB)	MS (AB) / Ec
Error (c) : Ec	(r-1) (p-1) (q-1)	Ec	
Total	**rpq - 1**		

Q.8. What is the null hypothesis and test statistic to be used to test the significance of the factor A.

Answer

$H_0 : a_j = 0$ i.e. there is no significant difference between the two (main effects) or treatment means of the factor A averaged over all the q levels of the second factor B.

The test statistic is

$$F = \frac{MSA}{Ea} \sim F_{(p-1),(r-1)(p-1)}$$

Q.9. Write the null hypothesis and test statistic to be used to test the significance of different levels of the factor B.

Answer

$H_0 : b_k = 0$ i.e. there is no significant difference between the two (main effect) or treatment means of the factor B. Averaged over all the p levels of the factor A

The test statistic is

$$F = \frac{MSB}{Eb} \sim F_{(q-1),(r-1)(q-1)}$$

Q.10. Write the null hypothesis to test the significance of interaction effect AB and the distribution of F statistic to be used.

Answer

$H_0 : (ab)_{jk} = 0$ i.e. the two factors A and B have no interdependence.

The test statistic is

$$F = \frac{MS(AB)}{Ec} \sim F_{(p-1)(q-1),(r-1)(p-1)(q-1)}$$

Q.11. Write the estimate of S. E. of difference between two main effects of the factor A.

Answer

Estimate S E (Difference between two main effects of the factor A)

$$= \sqrt{\frac{2Ea}{rq}}$$

Q.12. What is the estimate of S. E. of difference between two main effects of the factor B.?

Answer

$\widehat{SE}$ (Difference between two main effects of the factor B)

$$= \sqrt{\frac{2Eb}{rp}}$$

Q.13. What is expression for estimating the S. E. of difference between two A means at the same level of B ?

Answer

$\widehat{SE}$ (Difference between two A means at the same level of B)

$$= \sqrt{2\left[\frac{Ea + (q-1)Ec}{rq}\right]}$$

Q.14. Write the estimate of S. E. of difference between two B means at the same level of A.

Answer

$\widehat{SE}$ (Difference between two B means at the same level of A)

$$= \sqrt{2\left[\frac{Eb + (p-1)Ec}{rp}\right]}$$

Q.15. It is observed that the three errors are of different magnitudes. Which one will be the smallest.

Answer

The error variance corresponding to the smaller plot size i.e. the error E_c will be smallest among E_a, E_b and E_c which is usually expected.

Q.16. Out of the two main effects A and B and interaction AB, which one will be tested most precisely and why ?

Answer

The interaction effect AB will be tested most precisely in virtue of the large number of d.f. available for the corresponding error.

Q.17. In strip plot design the precision is sacrificed on the main effects of both the factors A and B. What is the use of such type of sacrifice in estimating the interaction effect AB ?

Answer

The sacrifice in precision of the main effects of both the factors A and B provides higher precision on interaction which will be generally more accurately determined than in either *(i)* RBD or *(ii)* simple split plot design

Q.18. What is the demerit of strip plot design as compared with split plot design regarding estimation of error varainces applicable to the main effects of both the factors A and B.

Answer

In case of strip plot design, the d.f. for estimating the error variances applicable to the main effects of both the factors A and B are likely to be smaller than that in case of split plot design and consequently this design is not recommended unless practical consideration necessitates to use it or unless the interactions are the principal objects of study.

Chapter 7

Progeny Row Trial and Compact Family Block Design

Q.1. What is progeny row trial ?

Answer

The replicated experiment designed to compare a number of parent plants with respect to any characteristic under study is known as progeny row trial.

Q.2. For what purpose the progeny row trial was investigated ?

Answer

In old days the mass selection technique was adopted in plant breeding to select the plants showing superior genetic values. Progeny row trial was investigated by plant breeders as an improved technique over mass selection technique.

Q.3. What is the characteristic feature of a progeny row trial ?

Answer

In progeny row trial each plot consists of a single row and all the seeds of which belong to the same parent plant. The trial is carried out in a simple RBD in which each plot is a progeny row and the various proganies to be tested are randomised in each blok.

Q.4. What is the structure of ANOVA table in progeny row trial ?

Answer

Let there be p progenies and r replications with k plants in each plot. The design is RBD therfore analysis of data will be same as that of RBD.

Source of variation	D.f.
Blocks or Replication	(r-1)
Progenies	p-1
Error (plot error)	(r-1) (p-1)
Pooled plant error	$\sum_{i}^{pr}(ki-1)$

Q.5. Give some examples of crops which are suitable for progeny row trial ?

Answer

Progeny row trial has been found to be suitable for wheat, cotton, sorghum, groundnut, limeseed etc.

Q.6. What is compact family block design ?

Answer

In the contrary of progeny row trial, if the progenies tried in an experiment belong to a number of strains, families etc. two types of comparisions are involved.

(i) The comparision between different familes

(ii) Comparision between progenies within families.

When it is intended to study the progeny differences within families, it is adivsed to use another design which utilizes the principle of local control more fully than a randomised block design with progenies randomised all over the block in order to reduce the error variance of progeny comparisions within families. Such a layout is known as compact family block design.

Q.7. What is randomisation procedure in compact family block design ?

Answer

In this design, the progenies belonging to the same family are sown side by side in a family main plot. The family plots are first randomised in each block and progeny plots which constitute the subplots, are randomised within the main plots.

Q.8. Is there any relation (similarity) between the compact family block design and split plot design ?

Answer

The compact famiily block design is quite analogus to the split plot design.

Q.9. For the comparision of the progenies within families, why it is advised to analyse the various families separately and to test the progeny differences within each family against the error obtained from the data of that family ?

Answer

For the comparison betwen progenies within family, the above type of advice is made because the error variance might differ significantly from family to family and consequently in taking a pooled estimate of error, the error might be over estimated in respect of some families and under estimated in other families.

Q.10. In case if the error variances do not differ significantly between the families even then it would not be appropriate to carry out an over all test of significance of differences between progenies within families. Why it is so ?

Answer

It is so because the progeny differences might be significant in some families and non significant in other families and latter might mask the former.

Q.11. Write the structure of pooled ANOVA table of compact family block design ?

Answer

Let there be f families, p progenies in each family. If there are r replications, the structure of ANOVA is as follows.

Variation due to	d.f.	M.S.	F ratio
Blocks (Replication)	r-1		
Families	f-1		
Error (a)	(r-1) (f-1)	Ea	
Total (Main Plot)	rf-1		
Progenies within family	f (p-1)		
Error (b)	fpr - 1	Eb	

Q.12. What is the difference between the compact family block design and the split plot design ?

Answer

In compact family block design, the subplot treatments are different for different main plots where as in the split plot design, the subplot treatments are same for all the main plots.

Q.13. What is the S. E. of difference between two progeny means belonging to two different families ?

Answer

S.E. (Difference between two progeny means taken from two different families)

$$= \sqrt{\frac{2}{r}\left[\frac{Ea+(p+1)Eb}{r}\right]}$$

Q.14. To test the homogenity of the error variance obtained from family to family, what test is useful ?

Answer

For this purpose the Bartlett test of homogenity of variances is useful.

Chapter **8**

Analysis of Covariance

Q.1. In what sense the analysis of covariance is an extension of ANOVA ?

Answer

The analysis of covariance is an extension of ANOVA in the sense that it covers the case where observations are taken on more than one variable from each experimental unit. The interest centers on the study variable y which depends upon on auxillary variable x.

Q.2. What type of test is of main interest in analysis of covariance?

Answer

The main interest in analysis of covariance is to test whether the variation in the dependent variable y over the classes is due to class effects or due to its dependence on the independent variable x which also vary from class to class.

Q.3. In what way the analysis of covariance controls the experimental error ?

Answer

The analysis of covariance controls the experiments error by taking into consideration of the dependence of y on x.

Q.4. What are the uses of the technique of analysis of covariance?

Answer

(1) To increase the precision of a randomized test experiment.

(2) To fit regression in a multiple classification.

Q.5. What are the assumptions required for valid use of the analysis of covariance ?

Answer

The assumption involving the regression term in the covariance are given below:

(1) The treatment and the regression effects may not be additive.

(2) The correct form of the regression line has been footed.

The danger of misleading results is greater when x shows real difference from treatment to treatment.

Q.6. What are the effects of violation of the assumptions made in the ANOVA ?

Answer

We made certain comments on the effects of violations of the underlying assumptions of

(1) Normality of the error and also the random effects.

(2) Independence and homosedasticity of errors.

In the fixed effects model, the normality assumption is needed only for hypothesis testing and interval estimation. Thus all the point estimators and their estimated variances remain valid even when normality does not hold. Heterosedasticity and correlation of errors do not bias the estimators. For other models too, the estimators of variance components remain unbiased even with non-normal random effects.

Q.7. What is the definition of analysis of covariance given by Prof. R.A. Fisher ?

Answer

In the words of Prof R. A. Fisher *"Analysis of covariance combines the advantage and reconciles the requirements of the two very widely applicable procedures known as regression and analysis of variance".*

Q.8. How it becomes possible to estimate the variation between blocks or between rows or between columns from the residual variation ?

Answer

It becomes possible by using the technique of local control.

Q.9. Write the definition of analysis of covariance ?

Answer

The very logical procedure which reduces the experimental error by estimating from it, the effects of variation in the auxiliary variate x and then increase the precision on the estimates of treatment means on the basis of the regression of study variate y on the auxiliary variate x, is known as the analysis of covariance.

Q.10. Give some examples of auxiliary variates if the study variate is yield of a crop.

Answer

Plant population, data of the uniformity trial, straw yield, number of tillers, age of crop. etc.

SECTION-B
SAMPLING DISTRIBUTIONS AND ASSOCIATED TESTS OF SIGNIFICANCE

Chapter 9

Tests Based on Large Sample (z Test)

Q.1. What do you mean by sampling distribution of a statistic ?

Answer

If we draw a sample of size *n* from a finite population of size N, then the total number of possible samples are

$$^{N}C_{n} = k\,(\text{say})$$

For each of these k samples we can compute some statistic $t = t\,(x_1, x_2, \ldots\ldots\ldots, x_n)$ eg. The sample mean $\overline{x}$ and variance s^2 etc. The set of values of statistic say $\overline{x} = (\overline{x}_1, \overline{x}_2, \ldots\ldots\ldots\ldots, \overline{x}_k)$ and $s^2 = (s_1^2, s_2^2, \ldots\ldots\ldots\ldots, s_k^2)$ constitute the distributions of the statistcs $\overline{x}$ and s^2. Such distribution of sample statistic is known as the sampling distribution.

Q.2. What do you mean by standard error of a statistic.

Answer

The standard deviation of the sampling distribution of a statistic is known as its standard error (S.E.) for example, in case of large sample, the sampling variance of the sample mean $\overline{x}_n$ is given by $\frac{\sigma^2}{n}$ *where* σ^2 is the population variance. Hence the

$$S.E(\overline{x}) = \sqrt{\frac{\sigma^2}{n}} = \frac{\sigma}{\sqrt{n}}$$

Q.3. What is the utility of the standard error of a statistic ?

Answer

The S.E. of any statistic plays a very important role in the large sample theory and forms the basis of the testing of hypothesis. If t is any statistic, then for large sample,

$$Z = \frac{t - E(t)}{S.E.(t)} \sim N(0,1)$$

Thus if $|t - E(t)| > Z\alpha.\ S.E.(t)$, the hypothesis is rejected at α% level of significance otherwise it is accepted at α% level signigicance.

Q.4. Is there any role of S. E. of any statistic in determining the probable limits within which the population parameter may be expected to lie ?

Answer

Yes, the S.E. of any statistic plays a very important role in determining the probable limits (confidence limits) for the population parameter within which, it is expected to lie. e.g. the sample mean $\overline{x}$ is the sample statistics to estimate the population mean μ, say, then the probable limit for μ is given by

$$(\overline{x} - 3.SE(\overline{x})) < \mu < (\overline{x} + 3.\ SE(\overline{x}))$$

$$\text{or } (\overline{x} - 3.\frac{\sigma}{\sqrt{n}}) < \mu < (\overline{x} + 3.\frac{\sigma}{\sqrt{n}})$$

Q.5. What are the significant values of z for a single tailed tests at commonly used levels of significance ?

Answer

The significant or critical value of z for a single tailes test (left or right) at the level of significance 'α' is same as the critical value of z for a two tailed test at the level of significance "2α".

The critical values of z at commonly used levels of significance are illustrated in the following table.

Critical values ($Z\alpha$)	**Level of significance (α) 1%**	**5%**	**10%**
Two tailed test	$\|Z\alpha\| = 2.50$	$\|Z\alpha\| = 1.96$	$\|Z\alpha\| = 1.645$
Right tailed test	$Z\alpha = 2.33$	$Z\alpha = 1.645$	$Z\alpha = 1.28$
Left tailed test	$Z\alpha = -2.33$	$Z\alpha = -1.645$	$Z\alpha = -1.28$

Q.6. If *n* is small, the sampling distribution of the test statistic *z* will not be normal then what will be normal, then what will be the alternative distribution of *z* ?

Answer

If the sample size *n* is small, then the test statistic *z* will follow the exact sampling distribution which may be any one out of t, F or χ^2 distributions. In general for $n < 30$, we consider the exact sampling distribution of the test statistic *z*.

Q.7. By comparing the calculated value and tabulated value of a test statistic, what is the rule to either reject or accept the null hypothesis ?

Answer

In the test of significance problem, the general rule is to reject the null hypothesis H_0 if the calculated value of test statistic is greater than the tabulated value of the corresponding test statistic at the level of significance 'α' otherwise it is accepted.

Q.8. What is the limit within which the value of standard normal variate z is expected to lie in all probability (α levels of significance) ?

Answer

The required limits within which the value of a standard normal variate z is expected to lie is given as $(-3 \leq Z \geq 3)$

Q.9. For a single tail test (right tail or left tail), what is the critical value (tabulated) $z\alpha$, at 5% level of significance with which the computed value of $|Z|$ is to be compared?

Answer

In case of single tail test (right tail or left tail test) the calculated mod value of Z i.e. $|Z|$ is compared with 1.645 which is the critical value of Z at 5% level of significance.

Test for significance for population mean

Q.10. What is an important assumption underlying the test of significance about the sampling distribution of the sample mean of a sample drawn from a population ?

Answer

A very important assumption underlying the test of significance is that the sample mean is asymptotically normally distributed even if the parent population from which the sample is drawn is not normal. Same is true though to a lesser extent of the sample standard deviation provided n is large.

Q.11. σ^2) *where* σ^2 is known. Then for a large sample, what will be the test statistic for testing the problem of single mean given as $H_0 : \mu = \mu_0$ vs $H_1 : \mu \neq \mu_0$

Answer

It is a two tailed test and the test statistic Z will be given as

$$Z = \frac{|\bar{x} - \mu_0|}{\frac{\sigma}{\sqrt{n}}} \sim N(0,1)$$

Q.12. If σ^2 is unknown then how it will be replaced in *z* statistic in case of large sample ?

Answer

In such a situation, for large sample size *n*, the z statistic will be defined by

$$Z = \frac{|\bar{x} - \mu_0|}{\frac{s}{\sqrt{n}}} \sim N(0,1)$$

For $s^2 = \frac{1}{n-1}\sum_i^n (x_i - \bar{x}_n)^2$ which is known as sample variance and is an unbaised estimate of σ^2.

Q.12*. Let $(\bar{x}_1, \bar{x}_2)$ be the means of the two independent samples of sizes (n1, n2) drawn from two normal populations N $(\mu_i, \sigma_i^2\ i = 1, 2)$, repectively, *where* $(\sigma_i^2\ i = 1, 2)$ are known. What will be the test statistics to test the hypothesis $H_0 : \mu_1 = \mu_2$ vs $H_1 : \mu_1 \neq \mu_2$

Answer

Here the problem is to test the equality of two normal populations and the appropriate test statistics is defined as

$$Z = \frac{|(\bar{x}_1 - \bar{x}_2) - E(\bar{x}_1 - \bar{x}_2)|}{SE(\bar{x}_1 - \bar{x}_2)} \sim N(0,1)$$

Which under H_0, reduces to the form

$$Z = \frac{|\bar{x}_1 - \bar{x}_2|}{\sqrt{\frac{\sigma_1^2}{n_1} + \frac{\sigma_2^2}{n_2}}} \sim N(0,1)$$

Q.13. In the problem discussed in question (12*), if $\sigma_1^2 \neq \sigma_2^2$ and are unknown, then in what way the Z statistic will be difined ?

Answer

In such situation, the Z statistic will be difined as

$$Z = \frac{|\bar{x}_1 - \bar{x}_2|}{\sqrt{\frac{s_1^2}{n_1} + \frac{s_2^2}{n_2}}} \sim N(0,1)$$

where $\hat{\sigma}_1^2 = s_1^2$, $\hat{\sigma}_2^2 = s_2^2$

This statistic results in some error, which is practically immaterial , if samples are large.

Q.14. If in the problem discussed in question (12*), $\sigma_1^2 = \sigma_2^2 = \sigma^2$ (say) but it is unknown. What will be the form of Z statistic to test $H_0 : \mu_1 = \mu_2$ vs $H_1 : \mu_1 \neq \mu_2$

Answer

In such a situation, the Z statistic will take the following form

$$Z = \frac{|\bar{x}_1 - \bar{x}_2|}{\hat{\sigma}\sqrt{\frac{1}{n_1} + \frac{1}{n_2}}} \sim N(0,1)$$

where $\hat{\sigma}^2 = \frac{(n_1 - 1) s_1^2 + (n_2 - 1) s_2^2}{(n_1 + n_2 - 2)}$

But in case of large sample sizes n_1 and n_2, in practice the following estimate without any serious error is used

$$\hat{\sigma}^2 = \frac{n_1 s_1^2 + n_2 s_2^2}{n_1 + n_2}$$

Test for significance of population proportion

Q.15. Let *x* be the number of success in *n* independent trials with constant probability P of success for each trial i.e. P is the population proportion and $p = \frac{x}{n}$ is sample proportion. What will be the test statistic for testing the $H_0 : p = P$ vs H_1 $p \neq P$

Answer

For the large sample size *n*, the normal test for the proportion of success will be given by

$$Z = \frac{|p - E(p)|}{SE(p)}$$

$$= \frac{|p - E(p)|}{\sqrt{\frac{PQ}{n}}} : N(0,1) \quad where\ Q = (1 - P)$$

Q.16. What are the probable limits for the observed proportion of success *p* (say).

Answer

The probable limit for the observed proportion of success p are given as

$$p - 3\sqrt{\frac{PQ}{n}} \le P \le p + 3\sqrt{\frac{PQ}{n}}$$

Q.17. If the population proportion P is not known, then what would be the probable limits for the proportion in population on the basis of sample proportion ?

Answer

The probable limits for P are given as

$$p \pm 3\sqrt{\frac{pq}{n}} \quad where\ q = (1 - p)$$

However the limits for P at level of significance α are given by

$$p \pm Z\alpha \sqrt{\frac{pq}{n}}$$

where $Z\alpha$ is the critical value of standard normal variate Z at the level of significance α.

Q.18. Let (X_1, X_2) be the number of persons possessing a given attribute A in random samples of sizes (n_1, n_2) drawn from two normal populations. Let population proportions be (p_1, p_2). Assuming (n_1, n_2) be large, what will be the test statistic for testing the $H_0 : P_1 = P_2 = P(say)$ vs $H_1 : P_1 \neq P_2$

Answer

Under the null hypothesis, the test statistic Z will be given as

$$Z = \frac{|p_1 - p_2|}{\sqrt{PQ\left(\frac{1}{n_1} + \frac{1}{n_2}\right)}} \sim N(0,1)$$

where Q = 1-P

Q.19. Suppose we want to test the significance of the difference between p_1 and p *where* $p = \frac{(n_1 p_1 + n_2 p_2)}{(n_1 + n_2)}$ is the pooled estimate of the population proportion. What will be the test statistics ?

Answer

In such a case, the test statistics is given by

$$Z = \frac{|p_1 - p|}{\sqrt{\frac{n_2}{(n_1 + n_2)} \frac{pq}{n_1}}} \sim N(0,1)$$

Q.20. Suppose the population proportions p_1 and p_2 are known and $p_1 \neq p_2$ and we want to test if the difference $(p_1 - p_2)$ is likely to be hidden in simple samples of sizes n_1 and n_2 drawn from two populations i.e. we want to test.

$H_0 : p_1 = p_2$ vs $H_1 : p_1 \neq p_2$

what would be the test statistic ?

Answer

In the usual notation, we have

$$Z = \frac{(p_1 - p_2) - E(p_1 - p_2)}{SE(p_1 - p_2)}$$

and under H_0, this test statistic becomes

$$|Z| = \frac{|P_1 - P_2|}{\sqrt{\frac{P_1 Q_1}{n_1} + \frac{P_2 Q_2}{n_2}}} \sim N(0,1)$$

Q.21. Let s_1 and s_2 be the standard deviations of two independent samples and we have to test the hypothesis.

$$H_0 : \sigma_1 = \sigma_2 \text{ vs } H_1 : \sigma_1 \neq \sigma_2$$

where σ_1 and σ_2 are unknown. Then for large samples, what will be the test statistic ?

Answer

In case of large samples drawn from two normal populations, the test statistic is given as

$$Z = \frac{|s_1 - s_2|}{\sqrt{\frac{\sigma_1^2}{2n_1} + \frac{\sigma_2^2}{2n_2}}} \sim N(0,1)$$

and since $\sigma_1^2 = \sigma_2^2$ are unknown, hence the test statistic reduces to the form

$$|Z| = \frac{|s_1 - s_2|}{\sqrt{\frac{s_1^2}{2n_1} + \frac{s_2^2}{2n_2}}} \sim N(0,1)$$

Chapter 10

χ^2 Distribution and Its Applications

Q.1. What is (chi-square) variate ?

Answer

The square of a standard normal variate is known as a chi - square (χ^2) variate with 1. d.f. Thus if $X \sim N(\mu, \sigma^2)$ then

$Z^2 = \left(\frac{x-\mu}{\sigma}\right)^2$ is known as chi-square variate with 1. d.f. and

is denoted by χ_1^2

Q.2. Explain the additive law of χ^2 variate.

Answer

If χ_i^2 (i = 1, 2, ..., n) are n independent chi-square variates each with 1. df. then the variate χ^2 defined as

$$\chi^2 = \sum_1^n \chi_i^2$$

is said to follow the chi-square variate with n. d.f.

Q.3. What is the density functin of χ^2 distribution ?

Answer

The p d.f of χ^2 distribution is represented as

$$f(\chi^2) = \frac{1}{2^{\frac{n}{2}}\sqrt{\frac{n}{2}}}(\chi^2)^{\frac{n}{2}-1}.e^{\frac{-x^2}{2}}d\chi^2 \qquad o < \chi^2 < \infty \text{ with n. d.f.}$$

Q.4. In what way the χ^2 test is sometime considered to be a non parameteric method ?

Answer

Since the probability function f (χ^2) does not involve any parameter (e.g. μ, σ^2) of the parent population, the χ^2 test is sometimes considered to be a non parameteric test.

Q.5. If χ^2 is a chi-square variate with n d.f. then what is the distribution of the variate $Y = \frac{\chi^2}{2}$?

Answer

The variate $Y = \frac{\chi^2}{2}$ will follow the gamma distribution with parameter $\left(\frac{n}{2}\right)$.

Q.6. If $\overline{x}$ is the sample mean of a random sample of size n drawn from a normal population. N (μ, σ^2). What will be the distribution of a variate χ^2 defined as $\chi^2 = \left(\frac{\overline{x} - \mu}{\sigma/\sqrt{n}}\right)^2$?

Answer

The variate $\chi^2 = \left(\frac{\overline{x} - \mu}{\sigma/\sqrt{n}}\right)^2$ will follow the chi-square distribution with 1 d.f.

Q.7. In what way one can say that the normal distribution is a particular case of χ^2 distribution ?

Answer

If we put n = 1, in the density function of χ^2 distribution, then we have

$$f(\chi^2) = \frac{1}{\sqrt{2\pi}}.e\left(\frac{-\chi^2}{2}\right).dx, \quad -\infty < \chi^2 < \infty$$

Which is the density function of standard normal variate. Thus, the normal distribution becomes a particular case of χ^2 distribution in case of degree of freedom being unity.

Q.8. What do you mean by standard χ^2 variate?

Answer

We know that the mean and variance of χ^2 distribution with n d.f. are to n and 2n respectively.

The standardized form of χ^2 variate is given as

$$Z = \frac{\chi^2 - \text{mean}}{\text{sd.deviation}} = \frac{\chi^2 - n}{\sqrt{2n}} \sim N(0,1) \text{ for } n \to \infty$$

and is known as standard χ^2 variate.

Q.9. In what way χ^2 distribution tends to follow normal distribution for large d.f. ?

Answer

We know that the standard χ^2 variate Z is defined as

$Z = \frac{\chi^2 - n}{\sqrt{2n}}$ has for large d.f. i.e. $n \to \infty$, the m.g.f. is given as

$M_z(t) = e^{\frac{t^2}{2}}$ which is the m.g. f of a standard normal variate. In other words, the χ^2 distribution tends to normal distribution for $n \to \infty$.

Q.10. Why the significant values of χ^2 statistic have been tabulated till n = 30 only ?

Answer

In practice for $n \geq 30$, the approximation of χ^2 distribution to normal distribution is fairly good i.e. for $n \geq 30$ the significant value of χ^2 statistic is approximately same as the corresponding significant value of standard normal variate Z. That is why the significant values of χ^2 statistic need not to be tabulated beyond $n \geq 30$.

Q.11. What do you know about the symmetric property of the χ^2 distribution ?

Answer

Since Pearsons, coefficient of skewness is given as $\sqrt{\frac{2}{n}}$ which is >0 for $n \geq 1$, the χ^2 distribution is positively skewed distribution. Further skewness is proportional to $\frac{1}{\sqrt{n}}$, it rapidly tends to symmetric as the d.f. increase and consequently as $n \to \infty$ the χ^2 distribution tends to normal distribution.

Q.12. What is cochran theorem related to χ^2 distribution ?

Answer

Let $X_1, X_2, \ldots\ldots X_n$ be independently distributed as standard normal variates, i.e. as N (0, 1). Let $\sum_{i}^{n} X_i^2 = Q_1 + Q_2 + \ldots\ldots\ldots + Q_k$ *where* each Qi is a sum of squares of line γ combination of $X_1, X_2, \ldots\ldots X_n$ with ni degree of freedom. Then if $\sum_{i=1}^{K} ni = n$, the quantities $Q_1, Q_2, \ldots\ldots, Q_k$ are independent χ^2 variates with $n_1, n_2, \ldots\ldots, n_k$ df. respectively.

Q.13. How the shape of chi-square probability curve changes with the d.f. and values of χ^2?

Answer

Let n be the d.f. and p (χ^2) as the height of the curve measured on y axis. Then for $n \geq 1$, as χ^2 increases, p (χ^2) decreases monotonically and finally tends to zero as $\chi^2 \to \infty$. Thus for n >1, the χ^2 probability curve is positively skewed towards higher values of χ^2. Moreover χ^2 axis is an asymptotic to the curve as given in the figure below.

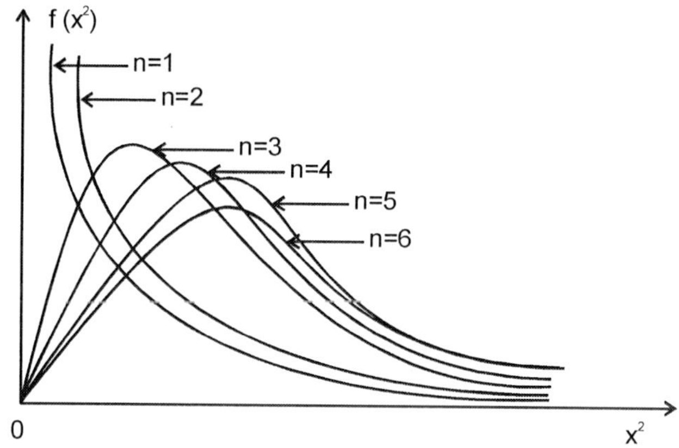

Q.14. If χ_1^2 and χ_2^2 are two independent χ^2 variates with n_1 and n_2 d.f. respectively, then what will be the distribution of new variate y defined as $y = \chi_1^2 / \chi_2^2$?

Answer

The variate y will follow the beta distribution given as $\beta_2\left(\frac{n_1}{2}, \frac{n_2}{2}\right)$

Q.15. Let χ_1^2 and χ_2^2 be two independent χ^2 variates with n_1 and n_2 d.f. respectively. The two new variates U and V are defined as

$$U = \frac{\chi_1^2}{\chi_1^2 + \chi_2^2} \text{ and } V = \chi_1^2 + \chi_2^2$$

What will be the distribution of U and V ?

Answer

The two variates U and V wll be independently distributed

as $U \sim \beta_1\left(\frac{n_1}{2}, \frac{n_2}{2}\right)$ and $V \sim \chi^2 \eta_1 + \eta_2$

Q.16. Let us consider a random sample of size n, whose members are distributed at random into K classes. Let pi be the probability that a sample observation will fall in the ith class (i = 1, 2, ……, K). Let ni (i= 1, 2, ….., K) be the observed frequency and npi be the corresponding expected frequency A statistic Y is defined as

$$Y = \sum_{i=1}^{k} \frac{(n_i - np_i)^2}{np_i}$$

What will be the distribution of *y* for a sufficient large *n*.

Answer

In case of sufficiently large *n*, the observed frequences ni (i = 1, 2, ….., k) will not be small and the statistic y defined as here will follow the χ^2 distribution with (k-1) d.f.

Q.17. For the validity of χ^2 test of "goodness of fit" between the theory and experiment, why a restriction is put that no theoretical cell frequency must be less than five ?

Answer

We know that the chi-square distribution is essentially a continuous distribution but can not maintain its character of continuity if any cell frequency is less than five. Hence to maintain the continuity character of χ^2 distribution, the restriction on cell frequency is imposed and a valid result of test of goodness of fit is obtained.

Q.18. How the critical value of χ^2 statistic & level of significance are affected with the change of degree of freedom of χ^2 ?

Answer

The critical values of χ^2 statistics increase with the increase of its degree of freedom but the level of significance decreases with the increase of in χ^2 values or in its d.f.

Q.19. Let $X_1, X_2, \ldots\ldots\ldots, X_n$ be a random sample from N (μ, σ^2). Then what will be the distribution of

$$\overline{x} = \frac{1}{n}\sum_{1}^{n} xi \;,\; \text{and} \;\; \frac{ns^2}{\sigma^2} = \sum_{i}^{n}\left(\frac{x_1 - \overline{x}}{\sigma}\right)^2$$

where $s^2 = \frac{1}{n}\sum_{1}^{n} (x_i - \overline{x})^2$

Answer

The two statistics $\overline{x}$ and $\frac{ns^2}{\sigma^2}$ are independently distributed

as $\overline{x} \sim N\left(\mu, \frac{\sigma^2}{n}\right)$ and $\frac{ns^2}{\sigma^2} \sim \chi^2\ (n-1)$

Q.20. If X and Y ar two independent random variates following the normal distributions as $N(\mu, \sigma_1^2)$ and $N(\mu_1, \sigma_2^2)$ respectively. A new variate ϕ is defined as

$\phi = \frac{(x - \mu_1)/\sigma_1}{(x - \mu_2)/\sigma_2}$ then what distribution will be followed by the random variate ϕ ?

Answer

The random variable ϕ being the ratio of two independent standard normal variates, is a standard cauchy variate with p d.f. given as

$$f(\phi, \theta) = \frac{1}{\pi}\frac{1}{1 + (\phi - \theta)^2}, -\infty \leq \phi \leq \infty$$

Q.21. Suppose we want to test if a random sample xi (i = 1, 2, ..., n) has been drawn from a normal distribution with a specified variance $\sigma^2 = \sigma_0^2$. Then what will be the test statistic ?

Answer

To test the null hypothesis $H_0 : \sigma^2 = \sigma_0^2$. The test statistic will

be given as $\chi^2 = \sum_{i=1}^{n}\left(\frac{x_i - \overline{x}}{\sigma_0}\right)^2$ and it will follow the chi-square distribution with (n-1) d.f.

Q.22. Let the two attributes (A, B) be divided into (r, s) classes respectively and are represented into (r, s) contigency table. Let $(Ai\ Bj)_o$ and $(Ai\ Bj)_E$ denote the observed and expected frequency of the persons possessing the attributes Ai and Bj. To test the hypothesis that the two attributes are independent, what will be the test statistic?

Answer

To test the independence of two attributes represented in the form of (r x s) contigency table, for large value of (N = r x s), the approximate test statistic (but fairly good) will be given as $\chi^2 = \sum_{i=1}^{r} \sum_{j=1}^{s} \left[\frac{\{(A_i \beta_j)_o - (A_i \beta_j)_E\}^2}{(A_i \beta_j)_E} \right]$ which will be distributed as χ^2 variate with (r-1)(s-1) d.f.

Q.23. To establish the closeness of the relationship between qualitative characters under study, Prof Karl Pearson suggested a measure known as "coefficient of mean square contingency" and denoted by C. How this measure C is defined ?

Answer

The coefficient of mean square contigency denoted by C is defined as $C = \sqrt{\frac{\chi^2}{\chi^2 + N}} = \sqrt{\frac{\phi 2}{1+\phi^2}}$ *where* $\phi^2 = \chi^2/N$ is known as mean square contingency.

Q.24. The coefficients of mean square contingency obtained for different type of classification are not comparable. Why it is so ?

Answer

For r x r contigency table, the maximum value of C is given by $\sqrt{\frac{r-1}{r}}$ which shows that the maximum value of C depends on r and s i.e. the maximum value of C differs for different classifications. Hence strictly speaking, the values of C obtained from different types of classifications inspite of being unitless absolute values are not comparable.

Q.25. In case of 2 X 2 contigency table, if any one of the theoretical cell frequency is less than S, then why the pooling method of χ^2 test is not recommended ?

Answer

In case of 2 X 2 contigency table, d.f. is equal to (2-1) (2-1) =1 and if the pooling method is applied then the d.f reduces to zero, which is meaningless. This is why the pooling method is not recommended.

Q.26. In a 2 X 2 contigency table, why the Yate's correction is recommended ?

Answer

In a 2 X 2 contingency table, to maintain the continuity character of the chi-square distribution, the yate's correction is recommended. This correction consists of adding 0.5 to those digonal cell frequencies whose multiplication is less than that of other digonal cell frequencies and 0.5 is substracted from the remaining cell frequency to make the marginal total of frequencies to remain same (constant).

Q.27. In case of a 2 X 2 contingency table, why yate's correction is recommended in case even if no cell frequency is less than five ?

Answer

In a 2 X 2 contingency table if no theortical cell frequency is less than five, even than the yate's correction is recommended in every 2 X 2 table only to maintain the continuity character of chi-square distribution.

Q.28. Let ri, (i = 1, 2, ., K) be K estiamtes of correlation coefficients from independent samples of sizes ni (i = 1, 2, ..., K) respectively. We want to test the hypothesis that these sample correlation coefficients are the estimates of the same correlation coefficient from a normal population. What will be the test statistic ?

Answer

This is the problem of testing the homogeneity of correlation coefficients and a chi-square statistic will be used to deal this problem which is defined as

$$\chi^2 = \sum_{i=1}^{k} (n_i - 3)(z_i - \overline{z})^2 \quad \text{with (k-1) d.f.}$$

where

$$Z_i = \frac{1}{2} \log_e \left(\frac{1 + ri}{1 - ri} \right), \qquad i = 1, 2, \ldots\ldots\ldots, k$$

$$\overline{Z} = \frac{\sum_{1}^{k} Z_i (n_i - 3)}{\sum_{1}^{k} (n_i - 3)}$$

Q.29. If the chi-square test results in the acceptance of equality of correlation coefficients then what will be the combined estimate of the population correlation coefficient ρ ?

Answer

In such situation ρ will be estimated as

$$\hat{P} = \frac{e^{2\overline{z}} - 1}{e^{2\overline{z}} + 1} = \tanh \overline{z}$$

Q.30. Let $r_1.s$, $r_2.s$, ... $r_k s$ be k estimates of partial correlation coefficient of order s_p from independent samples of sizes n_1, n_2, ..., n_k. respectively. We want to test the hypothesis that these sample partial correlation coefficients are the estimates of the sample partial correlation coefficient ρ.s of order s. from a normal population. What is the test statistic ?

Answer

To test the homogeneity of such partial correlation coefficients of order s, chi-square statistic is used which is defined as

$$\chi^2 = \sum_{i}^{k} (n_i - s - 3)(z_i - \overline{z})^2 \text{ with } (k-1)\,df.$$

where

$$Z_i = \frac{1}{2}\log_e\left(\frac{1+\gamma is}{1-\gamma i's}\right), \quad i = 1,2,\ldots\ldots\ldots,k$$

$$\overline{Z} = \frac{\sum_{1}^{k} Z_i (n_i - s - 3)}{\sum_{1}^{k} (n_i - s - 3)}$$

Q.31. Let $s_1^2, s_2^2, \ldots\ldots s_k^2$ be unbiased estimates of the population variance σ^2 obtained from k independent samples of sizes $n_1, n_2, .., nk$ respectively. We have to test the null hypothesis that the sample estimates s_i^2 (i=1, 2,.., K) of σ^2 are homogeneous. What is the test statistic and who has given it ?

Answer

The statistician Barttlett has proposed a chi-square statistic to test this hypothesis and is defined as

$$\chi^2 = \frac{\sum_{i=1}^{k} (ni \log \frac{s^2}{s_i^2})}{\left[1 + \frac{1}{3(k-1)}\left\{\sum_{1}^{k} \frac{1}{ni} - \frac{1}{n}\right\}\right]} \qquad \text{with } (k-1)\text{ df.}$$

where

$$ni = (ni - 1), \; n = \sum_{1}^{k} ni$$

$$s^2 = \frac{1}{n}\sum_{1}^{k} ni\, s_i^2$$

Q.32. If we have to test the null hypothesis discussed in question 31. In what way one can test the same hypothesis through F statistics without applying Bartlett test ?

Answer

Let s_i^2 and s_J^2, $i \neq j$, $1 \leq (i, J) \leq K$ be the smallest and largest values of the estimates of σ^2 respectively. If on the basis of F test, they do not differ significantly, then all the estimates s_i^2 $(i = 1, 2, ...k)$ can be reasonably regarded as homogeneous and coming from the same population ?

Q.33. What is the basic difference between the definitions of central and non-central chi-square distributions ?

Answer

The chi-square distribution defined as the sum of the squares of indpendent standard normal variates is often known as the central χ^2 distribution where as the sum of the squares of independent normal variates each having unit variance but with possibly non-zero mean is known as non-central χ^2 distribution.

Q.34. What do you mean by non-centrality parameter of a non-central χ^2 distribution ?

Answer

Let $X_i \sim N(\mu_i, 1), i = 1, 2, ...n$ be n independent normal variates. Then $\chi'^2 = \sum_i^k x_i^2$ has the non central χ^2 distribution with parameter (n, λ). where the parameter λ defined as $\lambda = \frac{1}{2}\sum_1^n \mu_i^2$ is known as the non-centrality parameter of the distribution.

Q.35. What do you mean by reproductive property of Non-central chi-square distribution ?

Answer

If Yi (i = 1, 2, ..., K) are independent non-central χ^2 variates with ni d.f. and non-centrality parameters λi then $Y = \sum_{I=1}^k Y_i$ is also a non-central X^2 variate with $\sum_{I=1}^k \eta_i$ d.f. and non-centrality parameter $\lambda = \sum_1^k \lambda_i$

Chapter 11

F and 't' Distributions and Their Applications

Q.1. Under what situation, the exact sample tests are required to be used and who discovered it first time ?

<u>Answer</u>

If the sample size n is small, the distribution of the statistic $z = \frac{\bar{x} - \mu}{\sigma/\sqrt{n}}$ if far from normality and as such normal test can not be applied if n is small. In such situations exact sample test, pioneered by W. S. Gossets, who wrote under the pen name of student and latter on, it was developed and extended by Prof. R. A. Fisher (1926).

Q.2. Give the names of some exact sample tests usually used in practice ?

<u>Answer</u>

The following are some tests which are known as exact sample tests based on exact sampling distribution.

1. Chi-square test

2. Student's t test
3. F test
4. Fisher's Z transformation

Q.3. Is it possible to apply the exact sample tests to the cases of large samples ?

Answer

Yes, the exact sample test can however, be applied to the large sample cases also but the converse is not possible.

Q.4. In all the exact sample tests, what is the basic assumption about the parent population from which the samples are drawn ?

Answer

In case of applying the exact sample tests, the basic assumption about the parent population from which the samples are drawn is that the parent population must be normally distributed.

Q.5. How the students t statistic is defined ?

Answer

Let xi, (i = 1, 2, ..., n) be a random sample of size n drawn from a normal population N(μ, σ^2). Then the students's t statistic is defined as $t = \frac{(\bar{x} - \mu)}{s/\sqrt{n}}$

where $\bar{x} = \frac{1}{n}\sum_{1}^{n} x_i$ is the sample mean and

$s^2 = \frac{1}{n-1}\sum_{1}^{n}(x_i - \bar{x})^2$ is an unbiased estimate of population variance σ^2. Thus t statistic follows the student's t distribution with (n-1) d.f.

Q.6. What is the form of student's 't' distribution ?

Answer

The probability density function of student's t distribution is given as

$$f(t) = \frac{1}{\sqrt{v}\beta\left(\frac{1}{2}, \frac{v}{2}\right)} \times \frac{1}{\left[1+\frac{t^2}{v}\right]^{\left(\frac{v+1}{2}\right)}}, \quad -\infty < t < \infty$$

Q.7. Under what condition, the student's t distribution reduces to cauchy distribution ?

Answer

If we put v = 1, in the density function of student's t statistic, we find that $f(t) = \frac{1}{\pi} \cdot \frac{1}{(1+t^2)}$ which is the p.d.f of standard cauchy distribution . Hence, when d.f. (v) of students's t distribution is put unity, the t distribution reduces to cauchy distribution.

Q.8. What is the original definition of student's 't' statistic defined as W. S. Gosset ?

Answer

W. S. Gosset originally defined the student's 't' in a slightly different way as $t = (\bar{x} - \mu)/s$ but latter on, it was extended by Prof. R. A. Fisher as $t = \frac{(\bar{x} - \mu)}{s/\sqrt{n}}$

Q.9. What is the definition of Fisher's 't' statistic ?

Answer

Fisher's t statistic is defined as the ratio of a standard normal variate to the square root of an independent mean chi-square variate. If ε is a N (0, 1) and χ^2 is an independent chi-square variate with n. d.f., then Fisher's 't' is given by $t = \varepsilon \Big/ \sqrt{\frac{\chi^2}{n}}$ and follows student's 't' distribution with n d.f.

Q.10. What do you mean by symmetric property of t distribution?

Answer

The t distribution is symmetrical about the line t = 0 and all odd order moments about origin vanish ?

Q.11. Let f (t) denotes the p d.f. of t distribution then how the f (t) varies with the variation in t ?

Answer

As t increases, f (t) decreases rapidly and tends to zero as t $t \to \infty$ so that x axis is an asymptotic to the curve.

Q.12. What is the limiting form of t distribution for $n \to \infty$

Answer

For $n \to \infty$, the limiting form of t distribution tends to standard normal variate.

Q.13. In comparative way, what do you know about the flatness of t distribution with that of normal distribution ?

Answer

The t distribution is more flat on the top than the normal curve.

Q.14. The significent values of t statistic at the level of significance 'α' and and d.f. have been tabulated for two tailed test only. If someone has to see the significant values of t statistic at the level of significance 'α' for a single tailed test (either left or right tailed) then what is the procedure to see the appropriate significant value ?

Answer

The significant values of t at the level of significance 'α' for a single tailed test can be obtained from those of two tailed test (given in the table) by looking the values at the level of significance '2α' For example

t_8 (0.05) for single tail test = t_8 (0.10) for two tail test

= 1.86

t_{15} (0.01) for single tail test = t_{15} (0.02) for two tail test

= 2.60

Q.15. What are the important assumptions for student's 't' test?

Answer

The following are the three important assumptions made for the student's 't' tests.

1. The parent population from which the sample is drawn is normal.
2. The sample observations are independent i.e. sample is random.
3. The population standard deviation σ is unknown

Q.16. Let $\overline{x}$ be the mean of a small sample of size n drawn from a normal population $N(\mu, \sigma^2)$ *where* σ^2 is unknown. What would be the test statistic to test the hypothesis that the sample mean $\overline{x}$ differs significantly from the hypothietical value μ_0 of the population mean μ ?

Answer

The appropriate test statistic will be the student's 't' statistic which is given as $t = \frac{|\overline{x} - \mu_0|}{s/\sqrt{n}}$ which follows the student's 't' distribution with (n-1) d.f. Here $s^2 = \frac{1}{n-1}\sum_{1}^{n}(x_i - \overline{x})^2$ is the unbiased estimate of σ^2.

Q.17. Let $(\overline{x}, \overline{y})$ be the sample means of two independent small samples of sizes (n_1, n_2) drawn from two normal populations $N(\mu_y, \sigma_x^2)$ and $N(\mu_y, \sigma_y^2)$ respectively. *where* $\sigma_x^2 = \sigma_y^2 = \sigma^2$ (say) (unknown). If we have to test the hypothesis that two samples have been drawn from two normal populations with the same means i.e. $\mu_x = \mu_y$ then what would be the test statistic ?

Answer

The approptiate test statistic would be the 't' statistic defined as:

$$t = \frac{|\bar{x} - \bar{y}|}{s.\sqrt{\frac{1}{n_1} + \frac{1}{n_2}}}$$

where

$$s^2 = \frac{\sum_{1}^{n_1}(x_i - \bar{x})^2 + \sum_{i}^{n_2}(y_i - \bar{y})^2}{(n_1 + n_2 - 2)}$$

and it will follow the student's t distribution with $(n_1 + n_2\text{-}2)$ df.

Q.18. If in the problem discussed in question 17, the two populations variances σ_x^2 and σ_y^2 are declared significantly different by F test then what alternative test will be applied ?

Answer

If the F test declares that $\sigma_x^2 \neq \sigma_y^2$ then in that case Behren's d test is applied which is complicated but a very satisfactory approximate statistic suggested by Cochran & Cox is applied which is as follows

$$t = \frac{|\bar{x} - \bar{y}|}{\sqrt{\frac{s_x^2}{n_1} + \frac{s_y^2}{n_2}}}$$

It is compared with the t* value defined as

$$t^* = \frac{t_1 v(\bar{x}) + t_2\, v(\bar{y})}{v(\bar{x}) + v(\bar{y})} = \frac{t.\frac{s_x^2}{n_1} + \frac{s_y^2}{n_2}}{\left(\frac{s_x^2}{n_1} + \frac{s_y^2}{n_2}\right)}$$

where t_1 : is the table value $t\alpha, (n_1\text{-}1)$

t_2 : is the table value $t\alpha, (n_2\text{-}1)$

and s_x^2 & s_y^2 have the usual meanings.

Q.19. Let us consider the following situations

(i) The sample sizes are equal i.e. $n_1 = n_2 = n$ (say)

(ii) The two samples are not independent but the sample observations are paired together i.e. the pair of observation. (xi, yi), (i = 1, 2, ..., n) corresponds to the same (i^{th}) sample unit. The problem is to test if the sample means differs significantly or not. What is the test statistic ?

Answer

In such a situation, we use the paired t test for significance of difference of means which is defined as

$$t = \frac{|\bar{d}|}{s/\sqrt{n}}$$

where

$$\bar{d} = \frac{1}{n}\sum_{1}^{n} di = \frac{1}{n}\sum_{1}^{n} (x_i - y_i)$$

$$s^2 = \frac{1}{(n-1)}\sum_{i}^{n} (d_i - \bar{d})^2$$

and this t statistic follows the student's distribution with (n-1) d.f.

Q.20. Let r be the observed sample correlation coefficient in a sample of *n* pairs of observations from bi-variate normal population. If we have to test the hypothesis that the population correlation coefficient *p* is zero what is the appropriate test statistic ?

Answer

To test the null hypothesis $H_0 : \rho = 0$, we use the following t statistic defined as

$$t = \frac{|r|\sqrt{n-2}}{\sqrt{1-r^2}}$$

which follows student's 't' distribution with (n-2) d.f.

Q.21. Let $r_{12 \cdot 34 \ldots (K+2)}$ be the partial correlation coefficient of order k, observed in a sample of size n drawn from a multivariate normal population having population partial correlation coefficient as $\rho_{12 \cdot 34 \ldots (K+2}$ To test the null hypothesis $H_0 : \rho_{12.34 \ldots\ldots (K+2)} = 0$

What is the appropriate test statistic ?

Answer

The null hypothesis H_0 will be tested with the use of following t statistic defined as

$$t = \frac{\left| r_{12.34 \ldots\ldots (k+2)} \right| \sqrt{n-k-2}}{\sqrt{1 - r^2_{12.34 \ldots\ldots (k+2)}}}$$

which follows the student's t distribution with (n-K-2) d.f.

Q.22. Let b be the sample regression coefficient of y on X obtained from a random sample $(x_i, y_i, i = 1, 2, \ldots, n)$ drawn from a bi-variate normal population having the population regression coefficient of y on x as β. We have to test the null hypothesis

$H_0 : b = \beta$ vs $H_1 \; b \neq \beta$

What is the appropriate test statistic ?

Answer

To test the hypothesis $H_0 : b = \beta$, we use the following t statistic defined as

$$t = \frac{|b - \beta| \sqrt{n-2 \sum_i^n (x_i - \bar{x})^2}}{\sqrt{\sum_1^n (y_i - \bar{y})^2 - b^2 \sum_1^n (x_i - \bar{x})^2}}$$

which follows the student's t distribution with (n-2) d.f.

Remark : In the problem considered in question number 22, if one has to test the hypothesis $H_0 : b = 0$ then the same statistic considered they will be used except putting $\beta = 0$.

Q.23. If someone considers the problem discussed in question 22, and want to test $H_0 : b = 0$ using F test or ANOVA technique then what is the test statistic ?

Answer

The appropriate test statistic F is given as

$$f_{(1, n-2)} = \frac{\text{MSS due to regression}}{\text{MSS from regression}}$$

$$= \frac{b^2 \sum_1^n (x_i - \bar{x})^2}{\frac{1}{(n-2)} \left[\sum_1^n (y_i - \bar{y})^2 - b^2 \sum (x_i - \bar{x})^2 \right]}$$

and will follow the F distribution with (1, n-2) d.f. Infact, this F statistic leads to the same conclusion as obtained in remark mention above.

Q.24. What is the definition of a non-central t distribution ?

Answer

Let X be a normal variate following N (μ, 1) and y be an independent χ^2 variate with n. df. Then non-central t statistic is defined as

$$t = \frac{X}{\sqrt{y/n}}$$

and is said to have a non-central t distribution with n d.f. and non-centrality parameter μ.

Q.25. How F statistic is defined ?

Answer

If χ_1^2 and χ_2^2 are two independent chi-square variates with v_1 and v_2 d.f. respectively, then F statistic is defined as

$$F = \frac{\chi_1^2 / v_1}{\chi_2^2 / v_2}$$

and is said to follow snedecor's F distribution with (v_1, v_2) d.f.

Q.26. Suppose we want to test the null hypothesis H_0 : two different normal population $N(\mu_x, \sigma_x^2)$ and $N(\mu_y, \sigma_y^2)$ have the same variance i.e. $\sigma_x^2 = \sigma_y^2$ How it will be tested ?

Answer

Let two independent samples xi, (i = 1, 2, ... n_1) and yj (j =1, 2, ...n_2) be drawn from $N(\mu_x, \sigma_x^2)$ and $N(\mu_x, \sigma_y^2)$ respectively. Let s_x^2 and s_y^2 be unbiased estimates of σ_x^2 and σ_y^2 respectively. Then the required test statistics is given by

$$F = \frac{s_x^2}{s_y^2}\ , \text{ if } s_x^2 > s_y^2\ , \quad = \frac{s_y^2}{s_x^2}\ , \text{ if } s_y^2 > s_x^2$$

and this F statistic follows the Sendecor's F distribution with (n_1-1, n_2-1) or (n_2-1, n_1-1) d.f. as case may be

Q.27. For what type of test (either one tail or two tail), the critical value of F distribution has been given in the table ?

Answer

In the table of F distribution, the critical value of F statistic has been tabulated for the right tail test only. Thus the significant value Fα (v_1, v_2) at the level of α and (v_1, v_2) d.f. is determined by

$$P\left[F > F_\alpha (v_1, v_2)\right] = \alpha$$

which is shown in the following diagram.

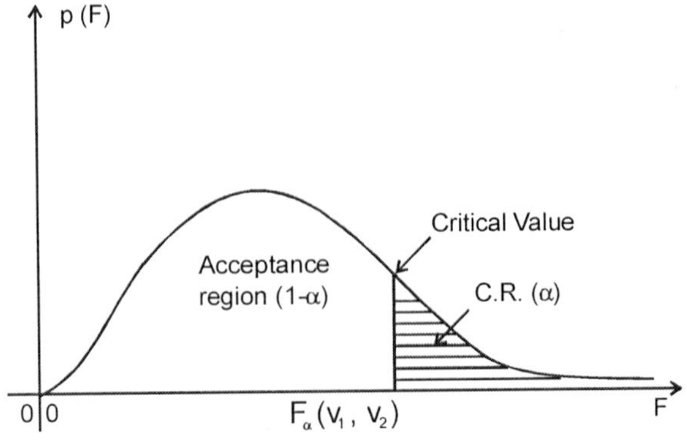

Q.28 What function relational exists between the students t distribution and Snedecor's F distribution ?

Answer

If a t statistic follows student's t distribution with n d.f., then t^2 follows Snedecor's F distribution with (1, n) d.f. Symbolically, if $t \sim t_n$

then $t^2 \sim F(1,n)$

Q.29. Can all the uses of t distribution be regarded as the applications of F distribution also ?

Answer

Using the relation $t_n^2 \sim F(1,n)$ all the uses of t distribution can be regarded as the applications of F distribution also e.g. to test for a single mean, instead of computing

$$t = \frac{|\overline{x} - \mu|}{s\sqrt{n}}$$

we may calculate $F = \dfrac{n(\overline{x} - \mu)^2}{s^2}$ and apply F test (1, n) d.f.

Q.30. What functional relation exist between F distribution and χ^2 distribution ?

Answer

In F (n_1, n_2) distribution, if we put $n_2 \rightarrow \infty$, then $\chi^2 = n_1 F$ and it follows χ^2 distribution with n1 d.f.

Q.31. Let R be the multiple correlation coefficient of a variate with k other variates in a random sample of size n drawn from a (K +1) variate population. Suppose someone wants to test the hypothesis

H_0 : The population multiple correlation coefficient. R = 0

What is the appropriate test statistic ?

Answer

We use the following F statistic

$$F = \frac{R^2}{1-R^2} \frac{n-k-1}{k}$$

which follows F distribution with (k, n-k-1) d.f.

Q.32. Let η be the observed sample correlation ratio obtained from a sample of size n from a bi-variate normal population arranged in *h* arrays. We have to test the hypothesis

H_0 : population correlation ration $\eta_{yx} = 0$

What is the appropriate test statistic ?

Answer

The appropriate test statistic is F statistic defined as

$$F = \frac{\eta^2}{(1-\eta^2)} \cdot \frac{(\eta-h)}{(h-1)}$$

which follows the F distribution with (h-1, n-h) d.f.

Q.33. Let η and r be the observed sample correlation ratio and correlation coefficient obtained from a sample of size n from a bi-variate normal population arranged in *h* arrays. We want to test the hypothesis

H_0 : The population regression is linear.

What is the appropriate test statistic for testing H_0 ?

Answer

The appropriate test statistic will be the F statisic defined as

$$F = \left(\frac{\eta^2 - r^2}{1-\eta^2}\right)\left(\frac{n-h}{h-2}\right)$$

and it follows the F distribution with (h-2, n-h) d.f.

Q.34. How a non-central F statistic is defined ?

Answer

Let χ_1^2 be a central chi-square statistic with n_1. d.f. and χ_2^2 be another independent non-central chi-square statisitc with n_2

d.f. and non-centrality parameter λ. Then the ratio F′ defined as

$$F^1 = \frac{(x_1^2/n_1)}{(x_2^2/n_2)}$$ is known as non-central F statistic and follows the F distribution with (n_1, n_2) d.f.

Q.35. What do you know about Fisher's Z distribution ?

Answer

In G. W. Snedecor's F distribution with (v_1, v_2) d.f., if we put

$$F = e^{2z} \Rightarrow Z = \frac{1}{2} \log e^F$$

and the distribution of Z is given by

$$g(z) = \frac{\left(\frac{V_1}{V_2}\right)^{\frac{V_1}{2}}}{\beta\left(\frac{V_1}{2}, \frac{V_2}{2}\right)} \cdot \frac{e^{V_1 Z}}{\left[1 + \frac{V_1}{V_2} e^{2Z}\right]} (V_1 + V_2)/2, \; -\infty < z < \infty$$

and is known as Fisher's Z distribution with (v_1, v_2) d.f.

Q.36. Who introduced the table of significant values of F statistic corresponding to various d.f. (v_1, v_2) with α% level of significance and how ?

Answer

Basically Prof. R. A. Fisher (in 1925, statistical method for research workers) published the table of significant values Fα of F with (α = 0.01, 0.05) corresponding to various d.f. (v_1, v_2). From these table values, Snedecor (1934-38), using the relation $F = e^{2z}$, introduced the table of significant values $F_\alpha (v_1, v_2)$ and denoted it by F in honour of Prof. R. A. Fisher.

Q.37. Can all the uses of F statistic be regarded as the applications of z statistic (distribution) ? If yes then how ?

Answer

Yes, all the uses of F distribution (statistic) can be regarded as the application of z distribution using the functional relation $F = e^{2z}$.

Q.38. In the limiting situation $v_1 \to \infty$, $v_2 \to \infty$, what will be the limiting situation of z distribution and what will be the mean and variance of the limiting distribution ?

Answer

In the limiting situation $v_1 \to \infty$, $v_2 \to \infty$ the z distribution tends to normal distribution with mean $\mu = \frac{1}{2}\left(\frac{1}{V_2} - \frac{1}{V_1}\right)$ and variance $\sigma^2 = \frac{1}{2}\left(\frac{1}{V_1} + \frac{1}{V_2}\right)$

Q.39. In a bi-variate normal population if the population correlation coefficient ρ is equal to ρ_0 ($\rho_0 \neq 0$), we want to test the hypothesis $H_0 : \rho = \rho_0$, then how it is tested using Fisher's z transformation ?

Answer

For testing the hypothesis $H_0 : \rho = \rho_0$. Prof. R.A.Fisher suggested the following transformation

$Z = \frac{1}{2} \log_e \frac{1+r}{1-r}$ where r is the sample correlation coefficient and proved that even for small n, the distribution of z is approximately normal with mean

$$\varepsilon_1 = \frac{1}{2} \log_e \frac{1+\rho_0}{1-\rho_0}$$

$$= 1/(n-3)$$

and variance (z) = 1/(n-3)

Hence the appropriate test statistic is given by

$$\phi = \frac{|z-\varepsilon|}{\sqrt{v(z)}} = (z-\varepsilon)\sqrt{n-3} \sim N(0,1)$$

Remark : This transformed z is not the same as standard normal variate z.

Q.40. Let r_1 and r_2 be the sample correlation coefficients observed in two independent samples of sizes n_1 and n_2 respectively. Suppose someone wants to test the following hypothesis

H_{01} : The sample correlation coefficients do not differ significantly or

H_{02} : The two samples have been drawn from the same population or

H_{03} : The two samples have been drawn from the two different populations having same correlation coefficient ρ, (say).

What would be the appropriate test statistic ?

Answer

Here all the three null hypothesis are same. Using the Fisher's z transformation, we use the z statistic defined as follows

$$\text{Let } Z_1 = \frac{1}{2}\log_e \frac{1+r_1}{1-r_1} \text{ and } Z_2 = \frac{1+r_2}{1-r_2}$$

and under the H_0, the appropriate test statistic is given as

$$Z = \frac{|z_1 - z_2|}{\sqrt{\left(\frac{1}{N_1-3}+\frac{1}{N_2-3}\right)}} \sim N(0,1)$$

Q.41. Let $r_1, r_2, \ldots\ldots\ldots\ldots, r_k$ be observed correlation coefficients in k independent samples of sizes $n_1, n_2, \ldots\ldots, n_k$ from a bi-variate normal population having population correlation coefficient as ρ. Then in what way one can combine these sample estimates to get a pooled estimate of ρ ?

Answer

We get the pooled estimate of ρ in the following way. Let $\hat{\rho}$ be the best estimate of ρ then we have

$$\hat{\rho} = \tan h \left[\frac{\sum_{1}^{K} (n_i - 3) z_i}{\sum_{i=1}^{K} (n_i - 3)} \right]$$

$$\textit{where} \quad Z_i = \frac{1}{2} \log_e \left(\frac{1 + r_i}{1 - r_i} \right), \; i = 1, 2, ..., k$$

SECTION-C
STATISTICAL INFERENCE

Chapter 12

Estimation Theory

Preliminaries

Q.1. What do you mean by statistical inference ?

Answer

The process of knowing all the relevant information from the known sample to the unknown populations is known as statistical inference.

Q.2. What is the problem of estimation ?

Answer

Some feature of the population in which an enquirer is interested, may be unknown to the inquirer and he may want to make a guess about this feature completely on the basis of a random sample drawn from the population. This type of problem is known as the problem of estimation.

Q.3. What are two different type of estimation problems ?

Answer

The point estimation and interval estimation.

Q.4. What is point estimation ?

Answer

Point Estimation : The estimation of a parameter by a single value (estimate of parameter) is known as point estimation.

e.g. The sample mean $\bar{y}_n$, sample correlation coefficient r_{xy} are the point estimates of the population mean μ and the population correlation coefficient ρxy respectively.

Q.5. What do you mean by interval estimation ?

Answer

The process to get an interval within which the population parameter is expected to lie, is known as interval estimation

e.g. The interval $\left(\bar{y}_n - t_{n-1,\alpha}\frac{s}{\sqrt{n}} < t_n < \bar{y}_n + t_{n-1,\alpha}\frac{s}{\sqrt{n}}\right)$ is taken as an interval estimate of the population mean μ.

Q.6. What do you mean by best estimator ?

Answer

An estimator is said to be best if it posses the properties of unbiasedness, consistency, efficiency and sufficiency.

Q.7. What do you mean by the unbiasedness property of an estimator ?

Answer

Any estimator whose mathematical expectation is equal to the population parameter, say θ, is known as an unbiased estimator for the parameter θ otherwise it is said to be biased estimator.

In mathematical notation an estimate t_n computed from sample values say $(x_1, x_2, ..., x_n)$ for a parameter θ, is said to be unbiased if

$E(t_n) = \theta$

e.g. the sample mean $\bar{y}_n$ gives an unbiased estimate of the population mean μ. In case of biased estimator, the amount of bias is computed from the expression

Bias $(t_n) = E(t_n) - \theta$

An estimator t_n is said to be positively biased or negatively biased for the parameter θ according as $E(t_n) > \theta$ or $E(t_n) < \theta$.

Q.8. What do you mean by an assymptotically unbiased estimator for the parameter θ if

$$\lim_{n \to \infty} E(t_n) = \theta$$

e.g. $t_n = \hat{\sigma}^2 = \frac{1}{n}\sum_{i=1}^{n}(x_i - \overline{x})^2$ is an assymptotically unbiased estimator for the population variance σ^2 since

$$\lim E\left(\frac{1}{n}\sum_{i=1}^{n}(x_i - \overline{x})^2\right) = \lim \frac{n-1}{n}\sigma^2$$

$$= \sigma^2$$

But $t_n^1 = \frac{1}{n}\sum_{1}^{n} x_i$ is absolutely unbiased estimator for population mean μ.

Remark : On unbiasedness.

1. If the sample mean $\overline{x}_n$ is an unbiased estimator for μ then note that any of the random variables $X_1, X_2, \ldots\ldots\ldots, X_n$ taken individually, say X_i, or only of the partial means say $\overline{x}_2 = \frac{1}{2}(x_1 + x_2)$ or $\overline{x}_3 = \frac{1}{3}(x_1 + x_2 + x_3)$, ... $X_{n-1} = \frac{1}{n-1}(x_1 + x_2 + \ldots\ldots + x_{n-1})$ is also unbiased. In other words we can state that the unbiasedness property of an estimator is independent of the sample size n.

2. The sample variance $s_1^2 = \frac{1}{n-1}\sum_{1}^{n}(x_i - x_n)^2$ and the partial variance say $s_2^2 = \frac{1}{n-2}\sum_{l=1}^{n-1}(x_l - \overline{x}_{n-1})^2$ etc. are also unbiased estimators of the population variacne σ^2.

Theoram 1. Let to be a M. V. U. estimator of the parameter θ and t_1 an unbiased estimator with efficiency e < 1. Then no unbiased linear combination of T_0 and T_1 can be a M.V.U. estimator.

Q.9. What is your views on the invariance property of an unbiased estimator ?

Answer

Let T be an unbiased estimator for $\gamma(\theta)$ and $\phi[\gamma(\theta)]$ be function of $\gamma(\theta)$ then $\phi(T)$ will not be unbiased for $\phi[\gamma(\theta)]$ unless $\phi(T)$ is a linear function of T even if T is an unbiased for $\gamma(\theta)$. In other words an unbiased estimator does not have an invariance property.

Minimum Variance Unbiased Estimator and Cramer, Rao Inequality

While an estimator may directly examine for unbiasedness, it is not immediately apparent how to satisfy oneself that an estimator has the smallest variance among all unbiased estimators.

The methods are available to achieve this task. One method is based on the use of the C. R. Inequality which it appears was presented independently and simultaneously by Rao, Cramer, Darmois and Frechet.

Q.10. What are the regularity conditions under which Cramer, Rao's inequality holds ?

Answer

The following are the regularity conditions

(i) The parameter t lies in an open interval of real line R.

(ii) $\frac{\partial}{\partial\theta} f(x,\theta)$ exists $\forall \theta \in \Omega\theta$

(iii) The range of integration is independent of the parameter θ so that $\int \prod_{i=1}^{n} f(x_i, \theta)\, d\underset{\sim}{x}$ can be differentiable under integral sign.

(iv) $E\left[\frac{\partial}{\partial\theta}\log f(x,\theta)\right]^2 > 0 \;\forall\; \theta \in \Omega\,\theta$

(v) $\int t_n(\underset{\sim}{x}).\prod_1^n f(x,\theta)\,d\underset{\sim}{x}$ can be differentiable under integral sign.

Q.11. What is the statement of C. R. 's inequality ?

Answer

Let $t_n(\underset{\sim}{x})$ be an unbiased estimator of a function $\psi(\theta)$ of the parameter θ then

$$V(t_n) \geq \frac{\left(\frac{\partial}{\partial\theta}\psi(\theta)\right)^2}{E\left(\frac{\partial}{\partial\theta}\log L\right)^2} = \frac{(\psi'(\theta))^2}{I(\theta)}$$

where I (θ) is the information on t suppied by the sample and

$$I(\theta) = E\left(\frac{\partial}{\partial\theta}\log L\right)^2$$

Q.12. What is your idea for C. R inequality to hold either for discrete or continuous distribution only ?

Answer

Cramer, Rao's inequality holds good for discrete as well as for continuous distribution.

Q.13. What is the procedure of finding a M. V. B estimator or M.V.U estimator for a parametric function ψ (θ) using C. R.'s inequality ?

Answer

If it is possible to write for any distribution, the partial derivative of the likelihood function in the form

$$\frac{\partial}{\partial\theta}\log L = \frac{1}{\lambda}[T-\psi(\theta)]$$

then using C.R. inequality, we find that T is an unbiased estimator for ψ (θ) with minimum variance λ i.e. T is M.V.B estimator or M.V.U. estimator for ψ (θ).

Q.14. In case of normal distribution N (μ, σ^2), what is the M. V. U. E for μ when σ^2 is known with its sampling variance ?

Answer

The sample mean $\overline{x}$ is a M. V. U. E. for μ with its sampling variance given as

$$V(\overline{x}) = \frac{\sigma^2}{n}$$

Q.15. In case of Poisson distribution P (x, λ), what is M. V. U. E. with its variance?

Answer

The sample mean $\overline{x}$ is a M. V. U. E for λ with its sampling variance given as $V(\overline{x}) = \frac{\lambda}{n}$

Q.16. What is M. V. U. E of p in B (n, p) with its sampling variance?

Answer

$$\text{M.V.U.E (p)} = \frac{r}{n}$$

$$\text{and } V(\text{M.V.U.E}) = \frac{pq}{n}$$

where r is the number of success in n trials.

Q.17. In case of cauchy distribution $f(x,\theta) = \frac{1}{\pi}\frac{1}{1+(x-\theta)^2}$ $-\infty \le x \le \infty$ what is the M. V. B estimator for θ

Answer

In case of cauchy distribution , no M. V. B. estimator exists for the parameter θ.

Q.18. What is M.V.B estimator with its sampling variance for the parameter σ^2 in case of normal distributor N (μ, σ^2) when μ is known.

Answer

The sample statistic $s_1^2 = \frac{1}{n}\sum_1^n (x_i - \mu)^2$ is M.V.B. estimator for σ^2

with minimum variance $V(s_1^2) = \frac{2\sigma^4}{n}$

Q.19. What do you mean by a regular estimator ?

Answer

It is an estimator for which certain regularity conditions hold, principally the differentiability of estimator w. r. to observation on which it depends and in case of frequency distribution, it should be differentiable w. r. to the parameters."

Q.20. What is the condition for the existance of M.V.B.E. ?

Answer

The equation $\frac{\partial}{\partial\theta}\log L = A(\theta)[T - \psi(\theta)]$

determines the condition on frequency distributions under which a M.V.B. estimaator for the parametric function $\psi(\theta)$ exists. where A (θ) is independent of observations but may be the function of θ.

Q.21. In your views, is there any difference between minimum variance bound estimator (M.V.B.E) and minimum variance unbiased estimator (M.V.U.E.) ?

Answer

Yes, An unbiased estimator T for ψ (θ) for which the C.R. lower bound is attained may be called as MVB estimator. Note that this is not the same as M.V.U estimator. For as we shall see that the least atttainable variance may be greater than the C.R. lower bound. Further in case of the regularity conditions underlying, this bound do not hold, the lest attainable variance may be less than the C. R. bound.

Q.22. If a M.V.B.E. exists for one specified function ψ (θ) of the parameter θ then does it exist also for the other function of θ ?

Answer

No, For example let us consider to find M.V.B.E. for σ^2 in case of N (μ, σ^2) where μ is known. The estimator $s_1^2 = \frac{1}{n}\sum_1^n (x_i - \mu)^2$ is M.V.B.E. for σ^2 with sampling variance

$V(s_1^2) = \frac{2\sigma^4}{n}$. But there is no M.V.B.E. for σ itself.

Q.23. What is the relationship for the coexistance of M.V.B.E and sufficient statistic for a parametric function $\psi(\theta)$ of the parameter ?

Answer

The M.V.B.E. for the parametric function (or parameter itself) exists only if there exists a sufficient statistic but converse may not be true.

Minimum Variance Unbiased Estimator (M.V.U.E.) and Black Wellisation

Q.24. What is the comperative usefulness of C. R. inequlity and Rao Blackwell theorem ?

Answer

The C.R. inequality provides us a means of judging whether or not a given unbiased estimator is also M.V.U.E. but Rao-Blackwell theorem enable us to obtain a M.V.U.E. from any other unbiased estimator by using a sufficient statistic say T for the parameter θ. The only condition to be statisfied is that T should have a complete family of distribution.

Q.25. What is your concept on the process Blackwellisation made by D. Blackwell ?

Answer

The implication of the Rao-Blackwell theorem is that if we are given an unbiased estimator U for the parameter θ then we can improve upon U by formig a new statistic ϕ (T) based upon U and a sufficient statistic T. This process of finding an improved estimator (from the point of view of unbiasedness and minimum variance) starting from an unbiased estimator has been called "Blackwellisation" after D. Blackwell.

Q.26. Under what condition, the estimator ϕ (T) will be better than U and even best estimator for the parameter μ ?

Answer

The estimator ϕ (T) will not only be a better estimator than U but also the best estimator in case if T has a complete family of distribution.

Q.27. In what sense the Rao - Blackwell theorem is less restructive than the C. R. inequality ?

Answer

The C. R. inequality method is too restrictive because of the regularity conditions imposed on it to hold good and also the fact that an estimator is not a M.V.B estimator does not mean that it is not a M. V. U. estimator for the parameter. Blackwellisation method does not suffer from this defects.

Q.28. How the Rao-Blackwell theorem is stated ?

Answer

Rao-Blackwell theoram : Let X and Y be two random variables such that

$E(Y) = \mu$ and $V(Y) = \sigma_y^2$

Let $E(Y/x = x) = \phi(x)$ Then

$E(\phi(x)) = \mu$ and $V(\phi(x)) \leq \sigma_y^2$

i.e. ϕ (x) is an unbiased estimator with minimum variance (best estimator) for the parameter μ. Here X is supposed to be a sufficient statistic having a complete family of distributions.

Q.29. What is a relationship between a M.V.B. estimator and a sufficient statistic ?

Answer

If a M.V.B. estimator exists, it is a function of sufficient statistic.

Results on UMVUE and MVUE

Q.30. In case of normal distribution N (μ, σ^2). What is UMVUE for σ^2 when .... ?

(i) μ is unknown

(ii) μ is known

Answer

(i) when μ is unknown then the statistic $s_1^2 = \frac{1}{n-1}\sum_1^n (x_i - \overline{x})^2$ is UMVUE for σ^2

(ii) when μ is known then the statistic $s_2^2 = \frac{1}{n}\sum_1^n (x_i - \mu)^2$ is UMVUE for σ^2

Q.31. In case of log normal distribution

$$f(x, \mu, \sigma^2) = \frac{1}{\sqrt{2\pi\sigma^2}} e - \frac{1}{2\sigma^2}(\log x - \mu)^2 \qquad x > 0 \quad \sigma^2 > 0$$

(a) what is UMVUE for μ

(b) what is UMVUE for σ^2 when μ is known

(c) what is UMVUE for σ^2 when μ is unknown.

Answer

(a) The statistic $T_1 = \frac{1}{n}\sum_1^n \log x_i$ is UMVUE for the parameter μ.

(b) In case μ is known,the statistic $T_2 = \frac{1}{n}\sum_1^n (\log x_i - \mu)^2$ is UMVUE for σ^2

(c) In case, when μ is unknown, the statistic

$$T_3 = \frac{n\sum^n (\log x_i)^2 - (\sum_1^n \log x_i)^2}{n(n-1)}$$ is UMVUE for σ^2

Q.32. In case of uniform distribution (continuous)

$f(x, \theta) = \frac{1}{\theta}$, $0 < x < \theta$, $\theta > 0$ what is UMVUE for θ and θ^r.

Answer

The statistics defined below as

$T_1 = \frac{n+1}{n} x_{(n)}$ and $T_2 = \frac{n+1}{n} x^{r}_{(n)}$ are the UMVUE of θ and θ^r respectively where $x_{(n)}$ is the max$(X_{(1)}, X_{(2)}, \ldots\ldots.., X_{(n)})$ of ordered set.

Q.33. In case of two parametric uniform distribution defined as

$$f(X, \theta_1, \theta_2) = \frac{1}{\theta_2 - \theta_1}, \quad \theta_1 < x < \theta_2$$

what are the UMVUE of θ_1 and θ_2?

Answer

The statistic T_1 defined as

(a) $T_1 = \frac{(n+1)X_{(1)} - \theta_2}{n}$ is UMVUE for θ_1 in case when θ_2 is known

(b) $T_2 = \frac{(n-1)X_{(n)} - \theta_1}{n}$ is UMVUE for θ_2 in case θ_1 is known

where

$X_{(1)}$ = Min $(X_{(1)}, X_{(2)}, \ldots\ldots\ldots, X_{(n)})$

$X_{(n)}$ = Max. $(X_{(1)}, X_{(2)}, \ldots\ldots\ldots, X_{(n)})$

Q.34. What is the UMVU estimator for θ in case of uniform distribution given as

$$f(X, \theta) = \frac{1}{2\theta}, \quad -\theta < x < \theta, \ \theta > 0$$

Answer

Let 'T = max $(- X_{(i)}, X_{(n)})$

Then $T' = \left(\frac{n+1}{n}\right) T$ is U.M.V.U. for θ.

Efficiency and Consistency

Efficient Estimator : If in the class of consistent estimators for a parameter θ, there exists one estimator whose sampling variance is minimum among such variances of other estimators of the class then it is known as " most efficient estimator". When such estmator exists, it provides the criteria for measurement of efficiency of other estimators.

The efficiency e of any estimator is defined as

$$e = \frac{\text{Variance of most efficient estimator}}{\text{variance of the given estimator}}$$

$e < 1.0$

Q.35. In estimating the mean of normal distribution, which one out of sample mean and sample median is more efficienct than other?

Answer

In case of normal distribution N (μ, σ^2), the sample mean and sample median are unbiased estimators for μ but the sample mean $\overline{x}$ is more efficient than the sample median $\tilde{x}$ since,

$$\text{Sampling variance (sample mean)} = \frac{\sigma^2}{n}$$

$$\text{Sampling variance (sample median)} = \frac{\pi}{2}\frac{\sigma^2}{n}$$

Q.36. In what sense, all the efficient estimators are said to be equivalent ?

Answer

All the efficient estimators are said to be equivalent in the sence that they are linearly related i.e. if T_1 and T_2 are two efficient estimators for the parameter θ then correlation coefficient (T_1, T_2) = 1.0.

Q.37. What do you mean by uniformally minimum variance estimator ?

Answer

An estimator ti ($\underset{\sim}{x}$) of θ_i is said to be uniformally minimum variance estimator for θ_i if $V(t_i(\underset{\sim}{x})) < V(h_i(x))$ for any other estimator h_i (x) of θi ∀ θi $\varepsilon \Omega_\theta$ where Ω_θ is the parameter space.

Consistency

Q.38. In what way the difference between the estimated value and the true value of a population parameter is described ?

Answer

Unbiasedness simply shows that the estimate on the aerage becomes equal to the parameter but it does not say anything about the difference between estimated value and the true value of the parameter. Consistancy describes this propety.

Q.39. What do you mean by the consistency property of an estimator ?

Answer

Let $\{T_n\}$ be a sequence of estimators for the parameter $\theta \in \Omega_\theta$ then the sequence $\{T_n\}$ is said to be consistent for θ if the sequence converge to θ in probability for every $\theta \in \Omega_\theta$. Thus Tn is said to be consistent if for every $\theta \in \Omega_\theta$ and for every pair of positive quantities $\in > 0, \eta > 0$ however small, there exists an no depending upon and n (and perhaps on θ) such that

$P_\theta\left[|Tn - \theta| \le \varepsilon\right] > 1 - \eta$ Whenever $n \ge n_0$

The above definition is analagous to convergence in mathematical sense. "In other word T_n is a consistent estimator for the parameter θ if it converges in probability to θ"

In other way, we consider the mean t_n of sample $x_1, x_2, ...x_n$ then if

$|t_n - \theta| \to 0$ as $n \to \infty$ i.e. if

$P\left[|t_n - \theta| > \varepsilon\right] \to 0$ as $n \to \infty$, then "t_n" is consistent estimator for θ.

Q.40. What is invariance property of a consistent estimator ?

Answer

Let T be a consistent estimator for $\gamma(\theta)$ and $\psi(\gamma(\theta))$ be a continuous function of $\gamma(\theta)$. Then $\psi(T)$ is a consistent estimator of $\psi(\gamma(\theta))$.

Remarks :

(i) A consistent estimator may not be an unbiased estimator and also an unbiased estimator may not be a consistent estimator but a consistent estimator with finite mean value may tend to be unbiased in large samples.

(ii) If there exists a consistent estimator tn, for θ, we can construct an infinite number of consistent estimators e.g. for fixed a, b, c, d, and e the estimators

$$\frac{an+b}{an+c}t_n, \frac{an^2+bn+c}{an^2+bn+d}t_n \text{ converges stochastically to } \theta$$

Q.41. What is sufficient condition for consistency ?

Answer

If T_n is an unbiased estimator for θ_n with variance σ_n^2 and $\theta_n \to \theta, \sigma_n^2 \to 0$ as $n \to \infty$ then T_n is consistent estimator for θ

More precisely if

$$\left.\begin{aligned} E(Tn) &= \theta n \to \theta \\ V(Tn) &= \sigma_n^2 \to 0 \end{aligned}\right\} \text{as } n \to \infty$$

then Tn is consistent estimator for θ.

Q.42. In case of normal distribution say, N (μ, σ^2), what are the consistent estimators for μ and σ^2 respectively ?

Answer

The sample median $\bar{x}$ and sample variance $s_1^2 = \frac{1}{n}\sum_1^n (x_i - \bar{x}_n)^2$ are the consistent estimators for the population mean μ and population variance σ^2 respectively.

Q.43. In cauchy distribution $f(x,\theta) = \frac{1}{\pi}\frac{1}{(1+(x-\theta)^2)}, -\infty < x < \infty$ what is the consistent estimators for parameter θ

Answer

In case of cauchy distribution, the sample mean is not a consistent estimator but the sample median $\tilde{x}$ is the consistent estimator for the parameter θ.

Sufficiency and Completeness

Q.44. What is the only information that guides an experimentor in making a decision about the value of the unknown parameter or parameters θ ?

Answer

The observed sample values $x_1, x_2, \ldots\ldots\ldots x_n$ are the only information which guide an experimentor in making a decision about the unknown parameter of the population.

Q.45. In most of the cases, the observations may be too numerous, too complicated and a set of numbers to be directly dealt with. In what way such type of data will be tackled to make some decision about the unknown parameter of the population ?

Answer

In such type of huge and complicated data, one should simplify the data by reducing (condensing) it through some relevant statistics to the parameter θ that sacrifice as little information as possible contained in the original data and this motive that leads to the notion of sufficient statistics.

Q.46. What is the principle of sufficiency (sufficient statistic)?

<u>**Answer**</u>

The principle of sufficiency is a principle for condensing or reducing the original observations $X_1, X_2 ...., X_n$ to a few statistics, which may then be used for the purpose of drawing inference about the probability distribution characterised by θ. Loosely speaking sufficiency refuction amounts to replacing $X_1, X_2, ... X_n$ by a few statistics, $T_1, T_2, ... T_k$ and thus discarding information which is not relevant to θ and retaining everything that is essential.

Preliminaries to the sufficient statistic : We shall suppose that the random variables $X_1, X_2, ..., X_n$ have the joint distribution function $F\theta$ which is known except for k (functionally unrelated) parameters $\theta_1, \theta_2.........., \theta_k$. We shall write $\theta = (\theta_1, \theta_2, ..., \theta_k)$ a vector with k components and shall suppose that the parameter space is Ω_θ. Consider K functionally unrelated statistics $T_1, T_2, ... T_k$, the whole set of which may be denoted by T.

Sufficient statistic (Definition 1) : $T_1, T_2, ..., T_k$ will be called sufficient statistics for θ (or rather for the family of distributions F_θ, $\theta \in \Omega_\theta$ if the conditional distribution of $X_1, X_2,$ ....X_n for any given set of a values of $T_1, T_2, ..., T_k$ is independent of θ.

Q.47. What is the condition imposed on random variables $X_1, X_2, ..., X_n$ under which the above definition holds good ?

<u>**Answer**</u>

The case where the random variables $X_1, X_2, ... , X_n$ are all discrete or all continuous, is concerned, the above definition holds good.

Remark : For the case of continuous random variables, the direct applications of the above definition (1) becomes difficult and to avoid such complication, we now give an alternative but equivalent definition which applies to discrete as well as to continous random variables.

Definition 2. Sufficient statistic : Consider together with $T_1, T_2, ..., T_k$, (n-k) additional statistics, say $T_{K+1}, T_{k+2}, ..., T_n$ which are functionally independent of other and also independent of $T_1, T_2, ..., T_k$. Then $T_1, T_2, ..., T_k$, are said to be sufficient for θ if the conditional distribution of $T_{K+1}, T_{K+2}, ..., T_n$ for given $T_1, T_2, ..., T_k$, be independent of θ whatever be the additional statistics.

Q.48. What is the defintiion of sufficient statistic given by Prof. R.A. Fisher ?

Answer

Let f (x, θ) be the p d.f. of random variable x with parameter θ. A statistic $T(\underset{\sim}{x}) = T(x_1, x_2, ..., x_n)$ is said to be sufficient statistic for the parameter θ if $T(\underset{\sim}{x})$ gives all the relevant information about θ.

Q.49. What is Fisher Neyman Factorisation theorem ?

Answer

Fisher Neyman Factorisation Theorem : Let $T(\underset{\sim}{x})$ be a sufficient statistic for θ with p d.f. denoted by g (T, θ). If the range of x is independent of the parameter θ then $T(\underset{\sim}{x})$ is sufficient for θ if.

$$\prod_{i=1}^{n} f(x_i,\theta) = g(T,\theta)\, h\,(x_1,x_2,......,x_n)$$

where $h(x_1, x_2, ..., x_n)$ is a function of $(x_1, x_2, ..., x_n)$ and is independent of θ.

Theorem 1 : A sufficient statistic is unique i.e. if T_1 and T_2 are two sufficient statistics for the same parameter θ then there exists a functional relationship between T_1 and T_2.

Theorem 2 : Let the statistic T be sufficient for the parameter θ and let ϕ (T) be a funtion in one to one correspondence with T then T′= ϕ (T) is sufficient for θ.

Theorem 3 : If $T(\underset{\sim}{x})$ is a sufficient statistic for θ then $T(\underset{\sim}{x})$ is sufficient for any function of θ.

Remark : In a much wider meaning of the term sufficiency, there always exists one set of sufficient statistics. For, the set of random variables $X_1, X_2, ..., X_n$ is itself sufficient for the parameter θ.

Again if, there are (absolutely) continuous (independently identically distributed random variables (i id.r.v's) then the set of random variables (order statistics) $X_{(1)}, X_{(2)}, ..., X_{(n)}$ also is sufficient for θ.

Distribution Possessing Sufficient Statistics

Case 1 : Range of variate is independent of parameter θ.

Q.50. When the range of variate is independent of parameter θ then what is the necessary condition for the existence of a sufficient statistic ?

Answer

Let f (x, θ) denotes the p d.f. of random variable (r.v.) X. Then we have

$$\frac{\partial}{\partial\theta}\log f(x,\theta) = p(\theta).k(x)+q(\theta)/n \qquad ....(A)$$

and it is the necessary condition for a sufficient statistic to exist for a parameter θ where p (θ) and q (θ) are the arbitrary functions of θ.

Q.51. What is the precise form of the exponential family of distributions which is the condition for the existence of a M.V.B estimator for some function of θ?

Answer

The required precise form of the exponential family of distributions is given as

$$f(x,\theta) = \exp\left[A(\theta).B(x)+D(\theta)+C(x)\right] \qquad ....(B)$$

and this result is due to Darmois, Pitman and Koopman.

Remark : If the above result (B) holds, it can be easily verified that if the range of f (x, θ) is independent of θ, the likelihood function yields a sufficient statistic for θ. Thus under this condition, the result given in (A), is sufficient for the distribution to posses the sufficient statistic.

Case 2 : Range of variate depends upon the parameters. We discuss form of the frequency distribution given below.

(i) When the lower terminal of the range depends upon the parameter θ, we then have the frequency function as

$f(x,\theta), \quad a(\theta) \leq x \leq b$

where a (θ) is monotone functions of x and θ is in some non-degenerate interval. Then the lowest order statistic x(i) can only be sufficient statistic for θ if f (x, θ) can be factorised as

$$f(x,\theta) = g(x)/h(\theta) \qquad \text{..............(C)}$$

where g (x) and h (θ) are the functions of x and θ respectively.

(ii) When the upper terminal of the range depends upon θ i.e. the frequency function is of the form

$f(x,\theta), \; a \leq x \leq b(\theta)$

where b (θ) is the monotone function of θ and θ is in some non degenerate interval then the highest order statistics $x_{(n)}$ is the single sufficient statistic for θ if (c) holds.

(iii) If both the terminal depends upon θ i.e. if the frequency function is fo the form

$f(x,\theta), \quad a(\theta) \leq x \leq b(\theta)$

and $L(x,\theta), = \prod_{i=1}^{n} f(x_i,\theta)\; U(X_{(1)} - a_{(\theta)})\; U(b_{(\theta)} - x_{(n)})$

Then the extreme obserations $X_{(1)}$ and $X_{(n)}$ are a pair of sufficient statistic for θ if

$f(x, \theta) = g(x)/\, h(\theta)$

Q.52. In case $f(x,\theta)$, $a(\theta) \leq x \leq b(\theta)$ and if a (θ) and b (θ) are monotone function in θ, does there exist a single sufficient statistic for θ or not ?

Answer

In such case, there does not exist a single sufficient statistic for θ but if it exists, it must clearly be a function of $X_{(1)}$ and $X_{(n)}$.

Remark : Essentially, one may ask whether a single sufficient statistic exists or not in case the range of random variable depends upon the parameter θ. It will tell us whether the product

U ($X_{(1)}$ - a (θ)) U (b (θ) - $X_{(n)}$) in L (x, θ) is equal to 1 or 0. It will only equal to 1 if both.

$x_{(1)} \geq a(\theta)$ and $b(\theta) \geq x_{(n)}$ hold

Then there are following possibilities

(a) Let f (x, θ) be factored as

$$f(x, \theta) = g(x) / h(\theta)$$

Given $f(x,\theta),\ a(\theta) \leq x \leq b(\theta)$ where a (θ) is monotone in θ there $X_{(1)}$ is single sufficient for θ and complete also for θ.

(b) Given $f(x,\theta),\ a(\theta) \leq x \leq b(\theta)$ where b (θ) is monotone in θ then $X_{(n)}$ is single sufficient and complete for θ.

(c) Given $f(x,\theta),\ a(\theta) \leq x \leq b(\theta)$ and if a (θ) and b (θ) are monotone in θ (i.e. both increasing or both decreasing function of θ then no single sufficient statistic exists but both $X_{(1)}$ & $X_{(n)}$ are jointly sufficient statistic for θ.

(d) Given $f(x,\theta),\ a(\theta) \leq x \leq b(\theta)$ and if a (θ) is a decreasing and b (θ) is an increasing function of θ. Then T = max {a^{-1} ($X_{(1)}$), b^{-1} ($X_{(n)}$)} is sufficient for θ and complete also for θ.

(e) Given $f(x,\theta),\ a(\theta) \leq x \leq b(\theta)$ and if a (θ) is an increasing function & b (t) is a decreasing function of θ then. T=min {a^{-1} ($X_{(1)}$), b^{-1} ($X_{(n)}$)} is sufficient and complete for θ.

Sufficient Statistic for Several Parameters

(a) Let there be a multivariate distribution rather than a univariate distribution. Thus if we define each X_i as a vector variate with ($p \geq 1$) components and t as the vector of statistic with s components.

If we can write

$$L(X, \theta) = g(t, \theta).\ h(x)$$

Then we call the components of t a set of (jointly) sufficient statistic for θ.

(b) We may have s as $s \leq k$ or $s \geq k$. If s = 1 we may call t as single sufficient statistic. If we put t = x, we see that the observations always constitute a set of sufficient statistic for θ with s = n.

(c) It evidently does not follow from the joint sufficiency of t for θ. That any particular component of t say $t^{(1)}$ is individually sufficient for θ_1. This will only be so if g (t, θ) factorizes with g_1 (t^1, θ) as one factor. Nor is the converse true always. The individual sufficiency of all the $t^{(i)}$ when others are known, does not imply joint sufficiency.

Q.53. In case of N (μ, σ^2) what is sufficient statistic for σ^2 when μ is unknown.

Answer

In case when μ is unknown, the statistic $s^2 = \frac{1}{n}\sum_{1}^{n}(x_i - \overline{x})^2$ alone is not sufficient statistic for σ^2 but the statistic $t = (\overline{x}\, s^2)$ is jointly sufficient for σ^2.

Corr. 1 : If μ is known then $s_1^2 = \frac{1}{n}\sum_{1}^{n}(x_i - \mu)^2$ alone is sufficient for σ^2

Q.54. In case of normal distribution N (μ, σ^2) what is sufficient statistic for μ when σ^2 is known ?

Answer

When σ^2 is known then the sample mean $\overline{x} = \frac{1}{n}\sum_{1}^{n} x_i$ is sufficient for μ.

Q.55. For normal distribution N (0, σ^2) what will be the sufficient for σ^2.

Answer

Since μ = 0, (known), the statistic $t = \frac{1}{n}\sum_{1}^{n} x_i^2$ sufficient for σ^2.

Q.56. Let $x_1, x_2, \ldots\ldots, x_n$ be a random sample from

$$f(x, t) = e^{-(x-\theta)}, \quad \theta < x < \infty$$
$$= 0 \text{ otherwise}$$

Then what is sufficient statistic for t.

Answer

Here $t = \min \{X_1, X_2, \ldots\ldots\ldots\ldots, X_n\}$ will be sufficient statistic for θ.

Completeness of a Sufficient Statistic

Q.57. When a sufficient statistic T is said to be complete ?

Answer

A sufficient statistic T is said to be complete if no function or it has zero expectation unless it is zero itself at most every where. i.e. T is said to be sufficient and complete if

$$E[\psi(T)] \Leftrightarrow P[\psi(T)=0]=1$$

Q.58. What do you mean by complete family of distributions ?

Answer

The family of distributions $\{F(T, \theta),\ \theta \in \Omega\theta\}$ is said to be complete if for any measurable function $\psi(T)$

$$E_\theta = [\psi(T)] = 0 \ \forall\ \theta \varepsilon \Omega_\theta$$
$$\Rightarrow \psi(T) = 0$$

almost everywhere for all $\theta \in \Omega_\theta$, *where* T is a sufficient statistic for θ. This concept of completeness makes use of point estimation as well as testing of hypothesis.

Q.59. What do you mean by boundedly completeness of a statistic or the family of distributions ?

Answer

A statistic T or the family of distributions $\{F(T, \theta),\ \theta \in \Omega_\theta\}$ is said to be boundedly complete if for any measurable function $\psi(T)$ such

$|\psi(T)| < M, \quad \text{for some } M$

and $E_\theta(\psi(T)) = 0 \ \forall\ \theta \in \Omega\theta$

$\psi(T) = 0 \ \forall\ \theta \in \Omega\theta$, almost every where

Some Important Results in Theorem Forms

Thm 1 : If T is a complete sufficient statistic for the parameter θ then it is necessarily boundedly complete but converse is not true.

Thm 2 : If T_1 and T_2 are jointly sufficient for two parameters θ_1 and θ_2 and further let T_1 be sufficient for θ_1 when θ_2 is known, then the conditional distribution of T_2 for a given value of T_1 is independent of θ_1.

Thm 3 : A sufficient estimator is most efficient estimator if most efficient estimator exists for the parameter θ.

Thm 4 : The method of minimum variance will give a solution only if there exists a sufficient statistic.

Thm 5 : If $T_1, T_2, \ldots\ldots\ldots, T_r$ is a set of jointly sufficient statistics then any set of one to one function or transformation of $T_1, T_2, \ldots\ldots T_r$ is also jointly sufficient.

Thm 6 : Let $X_1, X_2, \ldots\ldots\ldots, X_n$ be i id. rv's from a continous distribution f (x, θ) then the set of order statistic $X_{(1)}, X_{(2)}, \ldots\ldots\ldots X_{(n)}$ is sufficient for θ.

Thm 7 : The original sample drawn from a population is always a sufficient statistic for the parameter θ.

Minimal Sufficient Statistic

A sufficient statistic T is said to be minimal sufficient statistic if it is the function of every other sufficient statistic

or

A sufficient statistic T is minimal sufficient if the data as represented by the rv's $X_1, X_2, \ldots\ldots\ldots, X_n$ can not be reduced or condenced beyond T without sacrificing sufficiency.

The following theorems show some important results regarding minimal sufficiency and other sufficiencies.

Thm 1 : If f (x, θ) belongs to the regular case of one parameter exponential family of distributions and let T (xi) be sufficient statistic for t. Then $\sum_{1}^{n} T(x_i)$ is complete, minimal sufficient statistic for θ. If f (X, θ_1, θ_2, ..., θ_k) belongs to the regular case of k parameters exponential family then $\sum_{1}^{n} T_1(x_i)$, $\sum_{I=1}^{n} T_2(x_i)$,.............., $\sum_{1}^{n} T_k(x_i)$ are complete, minimal sufficient statistic for Q_1, Q_2, ..., Q_k.

Thm 2 : A consequence of the completeness of a statistic T is that only one function of that statistic can have a given expected value. Completeness, thus confirms a uniqueness property upon an estimator.

Thm 3 : Bounded completeness implies minimal sufficiency but the minimal sufficiency does not imply the bounded completeness.

Thm 4 : There can not be more than one boundedly complete sufficient statistic for the parameter θ.

Thm 5 : If T is boundedly complete sufficient statistic for θ and U is a minimal sufficient statistic for T then T is equivalent to U except possibly for a zero measure set.

Thm 6 : Let Xi (i = 1, 2, ..., n) and Yi (i = 1, 2,....., n) be independently normally distributed with same mean μ, and variances σ_1^2 and σ_2^2 respectively. Then $T = \{\bar{x}, s_1^2, \bar{y}, s_2^2\}$ is minimal sufficient but not boundedly complete where

$$s_1^2 = \frac{1}{n}\sum_{1}^{n}(x_i - \bar{x})^2 \text{ , } s_1^2 = \frac{1}{n}\sum(y_i - \bar{y})^2$$

Thm 7 : If a single sufficient statistic T exists for parameter θ then the M.L.E. $\hat{\theta}$ of θ is the function T alone.

Thm 8 : All the optimum properties of single sufficient statistic are conferred upon M.L. estimators which are one to one function of them.

Results on Complete Sufficient Statistic

(1) Binomial Distribution

$$f(x,p) = n_{c_x}\, p^x (1-p)^{n-x}, \qquad x = 0,1,2,\ldots\ldots..n$$
$$o < p < 1$$

Then

(i) Density function f (x, p) is complete w. r. to p.

(ii) $\sum_{l=1}^{n} x_i = r$ complete and sufficient statistic for the parameter p.

(2) Poisson Distribution

$$f(x,\lambda) = \frac{e^{-\lambda}\lambda^x}{x!}, \quad x = 0,1,2,\ldots\ldots\ldots..$$
$$\lambda > 0$$

Then

(i) The density function f (x, λ) is complete and sufficient w. r. to the parameter λ.

(ii) The statistic $T = \sum_{l=1}^{n} x_i$ is sufficient and complete for the parameter λ.

(3) Normal Distribution

$$f(x,\mu,\sigma^2) = \frac{1}{\sqrt{2\pi\sigma^2}} e^{-\frac{1}{2}\frac{(x-\mu)^2}{\sigma^2}}, \quad -\infty < x < \infty,\ -\infty < \mu < \infty,\ 0 < \sigma^2 < \infty$$

(i) The density function f (x, μ, σ^2) is complete wr. to μ where σ^2 is known but not even boundedly complete w.r. to σ^2 when μ. is known.

(ii) The statistic $T_i = \sum_{1}^{n} x_i$ is complete sufficient statistic for μ when σ^2 is known.

(iii) The pair of statistics $\left(\sum_{1}^{n} x_i, \sum_{i}^{n} x_1^2\right)$ is jointly complete and sufficient for (μ, σ^2).

(iv) $T = \sum_{i=1}^{n}(x_i - \mu)^2$ is complete sufficient statistic for σ^2 when μ is known.

(4) Log Normal Distribution

$$f(x, \mu, \sigma^2) = \frac{1}{\sqrt{2\pi\sigma^2}} e^{-\frac{1}{2}\frac{(\log x - \mu)^2}{\sigma^2}}, \quad x > 0, \sigma^2 > 0$$

(i) The density function f (x, μ, σ²) is complete wr. to μ when σ² is known.

(ii) The statistic $T_i = \sum_{1}^{n} \log x_i$ is complete and sufficient for μ when σ^2 is known.

(iii) The pair of statistic $\left(\sum_{1}^{n} \log x_i, \sum_{i}^{n} (\log x_i)^2\right)$ is jointly complete and sufficient for (μ, σ^2).

(iv) $T = \sum_{i=1}^{n}(\log x_i - \mu)^2$ is sufficient for σ^2 when μ is known.

(5) Uniform Distribution (Continuous)

(a) $f(x, \theta) = \frac{1}{\theta}, \quad 0 < x < \theta, \ \theta > 0$

Then

(i) Density function f (x, θ) is complete w.r. to θ

(ii) $T = \max(x_1, x_2, \ldots\ldots\ldots, x_n) = x_n$ is complete and sufficient for θ.

(b) $f(x, \theta) = 1, \quad \theta - \frac{1}{2} < x < \theta + \frac{1}{2}$

$$-\infty < \theta < \infty$$

Then the statistic $T = (x_{(1)}, x_{(n)})$ is sufficient but not even boundedly complete for θ.

(c) $f(x, \theta) = \frac{1}{2\theta}, \quad -\theta < x < \theta, \quad \theta > 0$

The statistic $T = \max(-x_{(1)}, x_{(n)})$ is complete and sufficient for θ.

(d) $f(x, \theta) = \frac{1}{\theta}, \quad k\theta \leq x \leq (k+1)\theta, \; k > 0$

The pair of statistic $T = (x_{(1)}, x_{(n)})$ is jointly sufficient for θ.

(e) Let $x_1, x_2, \ldots\ldots, x_n$ be a random sample from a double exponential distribution whose p d.f. may be taken as

$$f(x, \theta) = \frac{1}{2} e^{-|x-\theta|} \qquad -\infty < \theta < \infty$$

Here no single sufficient statistic exists for θ but the whole set of ordered observation

$T = (x_{(1)}, x_{(2)}, \ldots\ldots\ldots, x_{(n)})$ is sufficient for θ.

6. Two Parametric Distribution Function

$$f(x, \alpha, \beta) = \frac{1}{\beta - \alpha}, \qquad \alpha \leq x \leq \beta$$

Then

(i) $T_1 = \min\{x_{(1)}, x_{(2)}, \ldots\ldots\ldots, x_{(n)}\} = x_{(1)}$ is sufficient statistic for α when β is known.

(ii) $T_2 = \max\{x_{(1)}, x_{(2)}, \ldots\ldots\ldots, x_{(n)}\} = x_{(n)}$ is sufficient statistic for β when α is known.

(iii) The pair of statistic $T = (x_{(1)}, x_{(n)})$ is jointly sufficient statistic for (α, β).

Remark : When both the limits of x is the function of parameter θ then in that case, uniform distribution is not even boundedly complete.

Chapter 13

Methods of Estimation

Method of Maximum Likelihood Estimation (M.L.E.)

Q.1. What are the commonly used methods of estimation in practice ?

Answer

The following are the commonly used methods of estimation.

(i) Method of maximum likelihood estimation.

(ii) Method of minimum variance.

(iii) Method of moment

(iv) Method of minimum chi-square.

(v) Method of least square

(vi) Method of scoring.

Q.2. Which statistician initially formulated the method of maximum likelihood estimation but as a general method of estimation was first introduced by other ?

Answer

The method of M.L.E. was initially formulated by C.F. Gass but as a general method of estimation it was first time introduced by Prof. R.A. Fisher.

Q.3. What is likelihood function ?

Answer

Consider a random sample $x_1, x_2, \ldots\ldots, x_n$ from a distribution having p d.f. $f(x, \theta)$, $\theta \varepsilon \Omega_\theta$. The joint p d.f. of $x_1, x_2, \ldots\ldots, x_n$ is $L = \prod_{i=1}^{n} f(x_i, \theta)$. This joint p d.f may be regarded as the function of θ alone. When so regarded, it is called the likelihood function L of the random sample and we write

$$L(\theta, x_1, x_2, \ldots\ldots\ldots\ldots x_n) = L(\theta) = \prod_{i=1}^{n} f(x_i, \theta), \ \theta \varepsilon \Omega_\theta$$

Q.4. Why an estimator obtained through M. L. E method is known as maximum likelihood estimator ?

Answer

Let $\hat{\theta}$ be the M.L.E of the parameter θ and $L(\underset{\sim}{x}, \theta)$ be the likelihood function. If θ is replaced by $\hat{\theta}$ in L (x, θ) then the likelihood function L (x, $\hat{\theta}$) takes the maximum possible value i.e. L (x, $\hat{\theta}$) is atleast as great as L (x, θ) and hence $\hat{\theta}$ is known as maximum likelihood estimator of the parameter θ. In other words, the estimator $\hat{\theta}$ maximises the likelihood function.

Q.5. What are the equations which give the maximum likelihood estimator ?

Answer

The following equations give the solution for maximum likelihood estimator.

(i) $\frac{\partial}{\partial \theta} \log L(x, \theta) = 0$

(ii) $\frac{\partial^2}{\partial \theta^2} \log L(x, \theta) < 0$

and the equation $\frac{\partial}{\partial \theta} \log L(x, \theta) = O$ is usually referred as the likelihood equatiion.

Remark : The M.L.E. may or may not be the unique estimator for the parameter θ.

Q.6. What are the assumptions to be followed in case of method of maximum likelihood estimation ?

Answer

The following are the assumptions to be followed by the M.L.E. method

(i) The first and second order derivatives *viz*

$\frac{\partial}{\partial \theta} \log L(x,\theta)$ and $\frac{\partial^2}{\partial \theta^2} \log L(x,\theta)$ exist and are continuous function of θ in the range R (including true value θ_o of θ) for almost all x.

(ii) For every $\theta \in R$

$$\left| \frac{\partial}{\partial \theta} \log L(x,\theta) \right| < F_{1(x)} \text{ and } \left| \frac{\partial^2}{\partial \theta^2} \log L(x,\theta) \right| < F_{2(x)}$$

where F_1 (x) and F_2 (x) are integrable functions over $(-\infty, \infty)$

(iii) The 3rd order derivative $\frac{\partial^3}{\partial \theta^3} \log L(x,\theta) = \theta$ exists such that

$$\left| \frac{\partial^3}{\partial \theta^3} \log L(x,\theta) \right| > M_{(x)} \quad where \quad M_{(x)} < k$$

and M (x) is a positive quantity.

(iv) For every θ in R

$$E\left(-\frac{\partial^2}{\partial \theta^2} \log L(x,\theta) \right) = \int_{-\infty}^{\infty} \left(\frac{\partial^2}{\partial \theta^2} \log L(x,\theta) \right) d\underset{\sim}{x} = I(\theta)$$

where $0 < I(\theta) < \infty$ and is known as the information supplied by the sample.

(v) Range of integration is independent of parameter θ. But if the range of integration depends upon parameter t then f (x, θ) vanishes at extreme depending on θ. This assumption is made to make the differentiation under integral sign valid.

Q.7. What are the important properties followed by M.L.E. ?

Answer

The following are some important properties followed by M.L.E.

(i) Consistency

(ii) Assymptotic normality

(iii) Invariance

(iv) Efficiency under large sample

Q.8. What is your view on the existence of unique, and stationery or terminal M.L.E. for a parameter θ ?

Answer

The M.L.E. $\hat{\theta}$ is unique for θ if the likelihood function L (x, θ) has a unique maxima. The M.L.E. $\hat{\theta}$ is a stationery value (under regularity condition) or terminal value (maxima) according as the range of the variate is independent or dependent upon θ.

Q.9. What is Hazoor Bazar's Theorem ?

Answer

A consistent solution of the likelihood equation provides a maxima to likelihood function with probability tending unity as the sample size $n \rightarrow \infty$

Q.10. What do you mean by invariance property of a M.L.E. ?

Answer

Let T be a M.L.E. for γ (θ) and ψ (γ (θ)) be a continuous function of γ (θ). Then ψ (T) is M.L.E. of ψ (γ (θ)) and is known as invariance property of M.L.E. or consistent estimator for the parameter. e.g. since p (1-p) is a continuous function of p (binomial distribution), hence $\frac{r}{n}\left(1-\frac{r}{n}\right)$ is a consistent estimator for p (1-p) and likelihood $\frac{n}{r}$ is also a consistent estimator for $\frac{1}{p}$.

Remark : Even when the regularity conditions for the efficiency of M.L.E. do not hold, the M.L.E may have minimum variance property conferred upon it by the sufficient statistic of which the M.L.E. is always a function.

Q.11. Let T and U be consistent estimators of γ (θ) then what function of T and U will also be consistent estimator of γ (θ) ?

Answer

The function ϕ = a T + b U such that (a+b= 1) is also consistent estimator for r (θ).

Q.12. If we add to M.L. estimator, a function of the observation which tends in probability to zero then in what way the assymptotic property of the modified estimator will be affected ?

Answer

In such cases, the modified estimator will also have the assymptotic property.

Q.13. If a M.V.B estimator t exist for ψ (θ) and likelihood equation has a solution $\hat{\theta}$ then what relation will exist betweent and $\hat{\theta}$?

Answer

The relation established between t and θ will be given by $t = \psi(\hat{\theta})$ and the solution $\hat{\theta}$ is unique.

Q.14. Let $\{\hat{\theta}_n\}$ be the sequence of estimators obtained by L.F. then is the sequence $\{\hat{\theta}_n\}$ consistent ?

Answer

The sequence $\{\hat{\theta}_n\}$ will be a consistent sequence.

Q.15. In case of normal distribution N (μ, σ^2) what will be the M.L.E. of

(a) The population mean μ when σ^2 is known ?

(b) The population variance σ^2 when μ is known ?

(c) The simultaneous estimation of μ and σ^2 both ?

Answer

(a) When σ^2 is known then

$$\text{M.L.E.}(\mu) = \hat{\mu} = \overline{x}$$

(b) When the population mean μ is known then

$$\text{M.L.E.}(\sigma^2) = \hat{\sigma}^2 = \frac{1}{n}\sum_{1}^{n}(x_i - \mu)^2$$

(c) When σ^2 and μ, both are unknown then their simultaneous M.L.E.'s will be given as

$\hat{\mu} = \overline{x}_n$ (the sample mean)

$$\hat{\sigma}^2 = \frac{1}{n}\sum_{1}^{n}(x_i - \overline{x})^2$$

Q.16. What is M.L.E of λ in case of Poisson distribution

$$f(x,\lambda) = \frac{e^{-\lambda}\lambda^x}{x!}, \; x = 0, n...............$$

Answer

In case of Poisson distribution, the M.L.E. of λ is given as

$\hat{\lambda} = \overline{x}_n$ (The sample mean)

Q.17. What is the M.L.E of p in B (x, n, p)

Answer

$\hat{p} = \frac{r}{n}$ *where* r is the number of success of an event in *n* trials.

Note that $\left(\frac{r}{n}\right)^2$ is consistent for p^2.

Q.18. In case of the distribution

$$f(x,\theta) = \frac{1}{\sqrt{2\pi\theta}} e^{-\frac{1}{2}\left(\frac{x}{\theta}\right)^2}, \quad -\infty < x < \infty$$

what will be the M.L.E. of θ ?

Answer

$$\text{M.L.E. }(\theta) = \hat{\theta} = \sqrt{\frac{1}{n}\sum_{1}^{n} x_i^2}$$

Q.19. In case of rectangular population

$$f(x, \alpha, \beta) = \frac{1}{\beta - \alpha}, \quad \alpha < x < \beta$$

what is M.L.E's of α and β ?

Answer

Here the method of M.L.E. fails but M.L.E. is given as

$\hat{\beta} = X_{(n)}$ *where* $X_{(n)} = \max(x_1, x_2, \ldots\ldots\ldots x_n)$

$\hat{\alpha} = X_{(1)}$ $\quad X_{(1)} = \min(X_1, X_2, \ldots\ldots\ldots X_n)$

Q.20. In case of bi-variate normal distribution.

$$f(X, Y, \mu_1, \mu_2, \sigma_1, \sigma_2, \rho) = \frac{1}{2\pi\sigma_1\sigma_2(1-\rho^2)^{\frac{1}{2}}}$$

$$\text{Exp}\left\{-\frac{1}{2(1-\rho^2)}\left\{\left(\frac{x-\mu_1}{\sigma_1}\right)^2 - 2\rho\left(\frac{x-\mu_1}{\sigma_1}\right)\left(\frac{y-\mu_2}{\sigma_2}\right)^2\right\}\right\}$$

$$-\infty \le x, y \le \infty$$

$$0 < \sigma_1, \sigma_2 < \infty$$

$$-1 \le \rho \le 1$$

(a) What are the M.L.E. 's of ρ, σ_1^2 & σ_2^2 when μ_1 & μ_2 are known ?

(b) What will be the simultaneous M.L.E.'s of all five parameters $\mu_1, \mu_2, \sigma_1^2, \sigma_2^2$, & ρ. ?

Answer

(a) $$\hat{\sigma}_1^2 = \frac{1}{n}\sum_1^n (x_i - \mu_1)^2$$

$$\hat{\sigma}_2^2 = \frac{1}{n}\sum^n (y_i - \mu_2)^2$$

$$\hat{\rho} = \frac{\frac{1}{n}\sum_{i=1}^n (x_i - \mu_1)(y_i - \mu_2)}{\hat{\sigma}_1 \hat{\sigma}_2}$$

(b) In case of simultaneous estimation, we have

$$\hat{\mu}_1 = \frac{1}{n}\sum_i^n x_i = \overline{x}$$

$$\hat{\mu}_2 = \frac{1}{n}\sum_i^n y_i = \overline{y}$$

$$\hat{\sigma}_1^2 = \frac{1}{n}\sum_i^n (x_i - \overline{x})^2$$

$$\hat{\sigma}_2^2 = \frac{1}{n}\sum_i^n (y_i - \overline{y}_n)^2$$

$$\hat{\rho} = \frac{\frac{1}{n}\sum_i^n (x_i - \overline{x})(y_i - \overline{y})}{\hat{\sigma}_1 \hat{\sigma}_2}$$

Thus in case of simultaneous estimation, the M.L.E's of all the five parameters are the corresponding sample moments and the sample correlation coefficient.

Remark : Any unbiased estimator with finite mean is assymptotically consistent estimator for the parameter θ and any consistent estimator with finite mean is assymptotically unbiased for the parameter θ.

The merit of an estimator is judged by the distribution of the estimator i.e. by the property of sampling distribution of estimator.

Method of Minimum Variance : (Rao Blackwell Theorem)

The method of minimum variance will give a solution only if there exists a sufficient statistic (estimator) for the parameter θ.

Method of Moments

Q.21. Who initiated the method of moment for estimation of the parameters of a distribution ?

Answer

This method of estimation was first time initited by Prof. Karl Pearson.

Q.22. Which one out of the two methods say, the method of moments and method of maximum likelihood gives more efficient estimator for the parameter θ ?

Answer

In general the estimators given by the method of maximum likelihood estimation are more efficient than those obtained by the method of moment.

Q.23. Under what situation the method of moments and the method of M.L.E give identical solution for the parameter θ ?

Answer

Both the methods of estimation give the identical solution to estimate the parameter θ only if the parent distribution is of a particular form. In this case the ML estimators must be obtained as the linear functions of the moment.

Q.24. What is the most general form of the distribution for M.L.E. method and method of moments to give identical solution (estimator) for the parameter θ ?

Answer

The most general form as required is as $f(x, \theta) = \exp\{b_0 + b_1x + b_2X^2 + ...\}$. *where* b's depend upon θ's.

Q.25. $\chi_1, \chi_2, ..., \chi_n$ are i id random variables following distribution f (x, θ). Then what will be the consistent estimators for the population moments ?

Answer

In such a situation of i id random variables $\chi_1, \chi_2, ..., \chi_n$, the sample moments will be the consistent estimators for the corresponding population parameters.

Q.26. What do you know about the properties of assymptotic normality, efficiency and consistency of the estimators obtained through the method of moment ?

Answer

It has been shown that under quite general conditions, the estimators obtained by the method of moment are assymptotically normal but in general, they are not efficient as well as not consistent estimators.

Q.27. What is the most important practical use of method of moment?

Answer

The most important practical use of the method of moments is in the fitting of theoretical frequency curves to observe frequency distribution. Karl Pearson has used this method for estimating the parameters involved in fitting the so called Pearsonian curves.

Q.28. In what situation, the method of moment fails to give an estimate of the population parameter ?

Answer

The method of moment obviously fails to give an estimate of population parameter if the population moments do not exist as in case of cauchy distribution.

Method of Minimum χ^2

Q.29. What is the nature of observations to use the method of minimum χ^2 or modified minimum χ^2 method for estimating the population parameters ?

Answer

The method of minimum χ^2 and modified minimum χ^2 may be used when the observations are frequencies grouped into a finite number (say K) of mutually exclusive classes.

Q.30. In what situation the minimum χ^2 equation and likelihood equations are identical and produce the indentical estimates to the population parameters ?

Answer

In case when n, the number of observations is large, the minimum χ^2 equations and likelihood equations are identical, and provide the identical estimates for the population parameters.

Q.31. What do you mean by best assymptotic normal estimators ?

Answer

All those estimators which posses the assymptotic properties of the maximum likelihood estimators (i.e. consistency, efficiency and assymptotic normality) have been termed as best assymptotic normal (BAN) estimators.

Q.32. What are your views on possession of optimum properties by the least square estimators ?

Answer

The least square estimators do not have any optimum property even assymptotic property also.

Interval Estimation

Although the point estimates are quite useful, yet they leave something to be desired. In all those cases when the point estimator under consideration has a probability density function, the probability that the estimator actually equaled the value of the parameter being estimated is zero. (The probability that a continuous r.v. equals any one value is 0). Hence it seems desirable that a point estimate should be accompanied by some measure of the possible error of the estimate.

For instance, a point estimate might be accompained by some interval about the point estimate together with some measure of assurance that the true value of the parameter lies within an interval.

Q.33. What do you mean by interval estimation ?

Answer

When we consider a problem of estimation in such a way that a point estimate might be accompanied by some interval about the point estimate together with some measure of assurance that the true value of the parameter lies within the interval. Such type of estimation problem is known as interval estimation.

Q.34. What do you mean by confidence interval ?

Answer

Let xi (i = 1, 2, ..., n) be a random sample of size n from a population having p d.f. as f (x, θ), θ being parameter and we wish to find two statistic t_1 and t_2 as the function of xi (i=1, 2, ..., n) such that we can assume that the value of θ lies within an interval ($t_1 < \theta < t_2$) with certain degree of confidence i.e. $P (t_1 < \theta < t_2) = \alpha$ ... (1)

The interval ($t_1 < \theta < t_2$) is known as confidence interval or feducial limits for θ.

Q.35. What are your view on the randomness of the confidence interval ?

Answer

The confidence interval is always a random interval because it is based on the values of random samples.

Q.36. What do you mean by confidence coefficient ?

Answer

The confidence that an interval ($t_1 < \theta < t_2$) is correct in long run of statistical experience is measured by the probability α and this number ($0 < \alpha < 1$) is called the confidence coefficient.

Q.37. What do you mean by confidence belt or region of acceptance?

Answer

We know that the lower and upper limits i.e. the quantities t_1 and t_2 depend only on the α and n. For any fixed α, the totality of values of t_1 and t_2 from different samples determines a field within which θ is asserted to lie. This field is known as the confidence belt or region of acceptance.

Q.38. Let a confidence interval for the parameter μ be given as $p[\overline{x}-2.94<\mu<\overline{x}+2.94]=0.95$, then what is its meaning?

Answer

The meaning of the above result is that the probability that the random interval $(\overline{x}-2.94,\ \overline{x}+2.94)$ covers the unknown true mean μ is 0.95.

Q.39. What is your opinion about thc cxistence of so many possible intervals with the same probability for the parameter, say θ ?

Answer

In practice it has been observed that there are in fact, many possible intervals with the same probability (with the same confidence coefficient). For example we have

$$p[\overline{x}-2.94<\mu<\overline{x}+2.94]=0.95$$
$$=p[2.7-2.94<\mu<2.7+2.94]=0.95$$
$$=p[-0.24<\mu<5.64]=0.95 \text{ for } \overline{x}=2.7$$

and another 95% confidence interval for μ is given as (-1.35, 5.22)

Q.40. What do you mean by one sided confidence interval ?

Answer

Let xi (i = 1, 2, ..., n) be a random sample from the density f (x, θ). Let $T_1 = t_1(x_1, x_2, \ldots, x_n)$ be a statistic for which

$P[T_1 < t(\theta)] = \gamma$ then T_1 is called a one sided lower confidence limit for $t(\theta)$. Similarly, let $T_2 = t_1(x_1, x_2, \ldots, x_n)$ be a statistic for which $P[t(\theta) < T_2] = \gamma$ then T_2 is called one sided upper confidence interval for $t(\theta)$.

Remark : If a confidence interval for θ has been determined, then in essense, a whole family of confidence intervals has been determined. More specifically, for a given 100 γ percent confidence interval estimator of $t(\theta)$ can be obtained, where $t(\cdot)$ is any strictly monotone function. For example if $t(\cdot)$ is a monotone increasing fucntion and $T_1 = t_1(x_1, x_2, \ldots, x_n)$, $T_2 = t_2(x_1, x_2, \ldots, x_n)$ is a 100 γ parent confidence interval for θ, then $(t(T_1), t(T_2))$ is a 100 γ% confidence interval for $t(\theta)$ since $P\theta(t(T_1) < t(\theta) < t(T_2)) = P_\theta(T_1 < \theta < T_2) = \gamma$

Q.41. What do you mean by pivotal quantity ?

Answer

Let xi (i = 1, 2,.........., n) be a random sample from the density $f(x, \theta)$. Let $Q = q_1(x_1, x_2, \ldots, x_{n,\theta})$, that is let Q be a function of xi (i = 1, 2, ..., n). If Q has a distribution free from θ (i.e. independent of parameter θ), then Q is defined to be a pivotal quantity.

Q.42. Give an example of a pivotal quantity?

Answer

Let $x \sim N(\mu, \sigma^2)$. Then the quantity Z defined as

$Z = \left(\frac{x - \mu}{\sigma}\right)$ is a pivotal quantity because $Z \sim N(0,1)$ but another quantity Z, defined as $Z^1 = (\overline{x} - \mu)$ is not a pivotal quantity since $Z \sim N(0, \sigma^2)$

Q.43. In case of interval estimation, in what sense the confidence interval obtained is known as central or non-central ?

Answer

In case of two tailed test let α_1 and α_2 be the left and right tail area such that $\alpha_1 + \alpha_2 = \alpha$, the level of significance. If $\alpha_1 = \alpha_2$

then the confidence interval obtained in this case is known as central and in contrary cases, the intervals are called non central.

Q.44. Let $x \sim N(\theta, \sigma^2)$ where σ^2 is known. What is the expression for confidence interval for the parameter θ with confidence coefficient $(1-\alpha)$?

Answer

The required confidence interval for θ will be given by the expression.

$$\left(\bar{x} - Z_{\alpha/2}\frac{\sigma}{\sqrt{n}} \leq \theta \leq \bar{x} + Z_{\alpha/2}\frac{\sigma}{\sqrt{n}}\right)$$

where n is the sample size, $\bar{x}$ is the sample mean, α is the level of significence and $Z\alpha$ is the Z value at α% level of significance.

Q.45. Let $x \sim N(\mu, \sigma^2)$ where both the parameters μ and σ^2 are unknown. Write the expression to find the confidence limit for μ.

Answer

The required expression is given as

$$\left(\bar{x} - t_{\alpha/2,(n-1)}\frac{s}{\sqrt{n}} \leq \mu \leq \bar{x} + t_{\alpha/2,(n-1)}\frac{s}{\sqrt{n}}\right)$$

where t_α is the value of t statistic at α% level of significance for (n-1) d.f. and s^2 is the sample variance defined as

$$s^2 = \frac{1}{n-1}\sum_1^n (x_i - \bar{x})^2$$

Q.46. Let xi (i = 1, 2,........, n) be a random sample from $N(\mu, \sigma^2)$ *where* μ is known but not σ^2. What will be the confidential interval for σ^2 ?

Answer

The required confidential interval for σ^2 is given as

$$\left(\sum_{1}^{n}(x_i - \mu)/\chi^2_{\alpha/2,n-1} \leq \sigma^2 \leq \Sigma\,(xi - \mu)^2/\chi^2_{1-\alpha/2,n-1}\right)$$

where χ^2 is a chi-square value with n d.f.

Q.47. Let xi (i = 1, 2, ..., n) be a random sample from N (μ, σ^2), both the parameters being unknown. What will be the confidence interval for estimating the parameter σ^2 ?

Answer

The required confidence interval will be given as

$$\left(\sum_{1}^{n}(x_i - \overline{x})^2/\chi^2_{\alpha/2,n-1} \leq \sigma^2 \leq \sum_{1}^{n}(x_i - \overline{x})^2/\chi^2_{1-\alpha/2,n-1}\right)$$

$$\text{or}\left(\frac{(n-1)s^2}{\chi^2_{\alpha/2,n-1}} \leq \sigma^2 \leq \frac{(n-1)s^2}{\chi^2_{1-\alpha/2,n-1}}\right)$$

Q.48. What do you mean by a random interval for the parameter θ.

Answer

A random interval is a finite or infinite interval ($t_1 < \theta < t_2$) where at least one of the end points t_1 and t_2 is a random variable.

Q.49. What are the reasons for a confidence interval not being unique ?

Answer

Confidence intervals are not unique because

(1) Different statistics are available to obtain (1-α) confidence interval and

(2) Different multiple of the same statistic in a two sided confidence interval can be used.

Q.50. What are the main criteria for selecting the confidence interval?

Answer

The main criteria for selecting the confidence interval are as follows

1. Minimise the length $\delta = (T_1 - T_2)$ of confidence interval (T_1, T_2). A weakness of this criteria is that in some cases the δ may be a random variable

2. Minimise the expected length $E(T_2 - T_1)$. A weakness of this criteria is that the minimisation of $E(T_2 - T_1)$ depends upon the unknown true value of the parameter.

Q.51. In case of large samples, the confidence intervals depending upon what factor, have the minimum expected length ?

Answer

In case of large samples, the confidence interval have the minimum expected length when they are based on $\left(\frac{\partial}{\partial \theta} \log L\right)$ *where* L is the likelihood function.

Chapter 14

Theory of Testing of Hypothesis

Preliminaries and Test Based on N. P. Lemma

Q.1. What do you mean by problem of testing of hypothesis ?

Answer

Some feature of the population may be available to the scientist and he may want to see whether the information on feature is teneble in the light of random sample drawn from the population. Such type of problem is known as the problem of testing of hypothesis.

Q.2. What do you mean by a test of statistical hypothesis ?

Answer

A test of statistical hypothesis is a rule which, when the experimental sample values have been obtained, leads to a decision to accept or to reject the hypothesis under consideration.

Q.3. What is statistical hypothesis ?

Answer

A statement or an assertion about the distribution of one or more random variables, whose validity is sought to be examined on the basis of a set of observations on the random variable, is known as statistical hypothesis.

Q.4. What do you mean by null and alternative hypothesis ?

Answer

Any statistical hypothesis under consideration which is tested for possible rejection on the basis of information supplied by a random sample of size *n* drawn from the population f (X, θ) is referred to be the null hypothesis and the hypothesis which contradicts the null hypothesis is known as alternative hypothesis. These are denoted by H_0 : and H_1. eg.

$H_0 : \mu = \mu_0$

$H_1 : \mu \neq \mu_0$ for N (μ, σ^2)

Q.5. What is simple hypothesis ?

Answer

A statistical hypothesis which specifies a unique point for the parametric value in the parametric space $\Omega\theta$, is known as simple hypothesis, eg. If $x \sim N(\mu, \sigma^2)$. then

(1) $H_0 : \mu = 2, S^2 = 10$ and

(2) $H_0 \ \mu = 7, S^2 = 7.5$

are the example of simple hypothesis

Q.6. Give explanation about composite hypothesis ?

Answer

A statistical hypothesis which is composed of finite or infinite number of simple hypothesis known as composite hypothesis. The composite hypothesis leaves at least some of the population parameters unspecified and number of unspecified population parameters is termed as the degree of freedom of hypothesis. e.g. It $x \sim N(\mu, \sigma^2)$. then

(1) $H_0 : \mu = 10$

(2) $H_0 : \sigma^2 = 3$

(3) $H_0 : \mu < 10, \sigma^2 = 3$

(4) $H_0 : \mu = 10, \sigma^2 > 3$

are composite hypothesis.

Q.7. What is parameteric test ?

Answer

The statistical tests which deal with the parameters of the population (distribution) are termed as parameteric test.

Q.8. What do you mean by non-parametric test ?

Answer

The statisitcal tests which do not make any assumption about the form of distribution function of the population from which the sample is drawn, are termed as non-parametric test e.g. Sign test, Signed rank test, Wileoxon signed rank test. etc.

Q.9. What do you mean by significant event and non significant event ?

Answer

If the event is a very rare event and if it happens, it is termed as significant event while if the event is a very common event to happen and if it happens, then it is termed as non-significant event eg. if in the month of December, it rains very heavily and the sky remains cloudy for 5 or 6 days continuosly, then this event (rains) will be termed as significant event while if the same event happens in the month of July or August then it will be termed as non-significant event.

Q.10. What is level of significance ?

Answer

It is the specification of the probability level which is the deciding level to declare a statistic or estimate to be significant or non-significant. It is denoted by α which in practice generally has the value 10%, or 5% or 1%.

Q.11. What are the four possible situations which arise while dealing with a problem of testing of hypothesis..

Answer

At the stage of taking decision about the hypothesis, the following four types of decisions are possible out of which anyone can be finally taken.

(1) The null hypothesis is correct and it is accepted.

(2) The null hypothesis is correct and it is rejected.

(3) The null hypothesis is false and it is accepted

(4) The null hypothesis is false and it is rejected.

Q.12. What are the two different types of error committed while taking decision about the hypothesis ?

Answer

An error will be committed if one accepts the wrong hypothesis or rejects the correct hypothesis. Hence two types of errors are possible.

Type I error : If the null hypothesis H_0 is true but it is rejected then the error committed is termed as type I error and is denoted by α (the level of significance). This error is known as producer's risk.

Type II error : If the null hypothesis H_0 is false but it is accpeted then the error committed is termed as type II error and is denoted by β. This type of error is known as consumer's risk.

Q.13. Out of two types of errors, which one is considered more serious ?

Answer

Out of two types of errors committed, the type II error is considered more serious and one should try to avoid it. However, in cases, where risk is minor, it is sometimes tolerated.

Q.14. What is the region of rejection or critical region ?

Answer

Let X : (x_1, x_2, ..., x_n) be a simple random sample with sample space Rn. Let Rn be divided into two distinct regions C and A such that C + A = Rn. Then the region $C \subset R_n$ is called critical region for the hypothesis

$$\left.\begin{array}{l} H_0 : \theta \in \Omega\theta \\ H_1 : \theta \notin \Omega\theta \end{array}\right\}$$

if the rule is to reject the H_0 whenever $(x_1, x_2, \ldots, x_n) \in C$ and accept H_0 for otherwise.

Q.15. What is the region of acceptance ?

Answer

The region AC R_n is termed as the region of acceptance for the hypothesis

$H_0 : \theta \in \Omega\theta_0$

$H_1 : \theta \notin \Omega\theta$

if the rule is to accept H_0 whenever $(x_1, x_2, \ldots, x_n) \in A$ and reject H_0 for otherwise.

Q.16. Does the acceptance of H_0 mean that H_0 has been proved to be true ?

Answer

No, the accpetance of H_0 based on $(x_1, x_2, \ldots, x_n)$ simply implies that as long as the random sample $(x_1, x_2, \ldots, x_n)$ is concerned, the null hypothesis H_0 is plausible hypothesis.

Q.17. Does the rejection of H_0 mean that H_0 has been disproved?

Answer

No. The rejection of H_0 merely means that H_0 does not look plausible in the light of observations of the random sample $(x_1, x_2, \ldots, x_n)$.

Q.18. What do you mean by one tailed test ?

Answer

A test of any statistical hypothesis where the alternative hypothesis is one tailed (right tailed or left tailed) is called a one tailed test. For example, a test for testing the mean of a population N (μ, σ^2)

$H_0 : \mu = \mu_0$

$H_1 : \mu > \mu_0$ (right tailed) or H1 : $\mu < \mu_0$ (left tailed) is a single tailed test. In the right tailed test. ($\mu > \mu_0$) the critical region lies entirely in the right tail of the sampling distribution of $\overline{x}$ and similarily in case of left tailed test.

Q.19. What do you mean by two tailed test ?

Answer

A test of statistical hypothesis where the alternative hypothesis is two tailed such as

$H_0 : \mu = \mu_0$ vs $H_1 : \mu \neq \mu_0$, ($\mu > \mu_0$ or $\mu < \mu_0$)

is known as two tailed test and in such a case the critical region lies in both the tailes of the probability curve of the test statistic.

Q.20. What do you mean by critical value or significant values?

Answer

The value of the test statistic which separates the the critical region and the acceptance region is called the critical value or significant values. It depends upon

(1) The level of significance used.

(2) The alternative hypothesis, whether it is two tailed or single tailed.

Q.21. What do you mean by the power of the test ?

Answer

The probability of rejecting the null hypothesis H_0 when infact it is false, is called the power of the test or power of the critical region. It measures in a sense the safeguard provided by the test against accepting a false hypothesis. It is measured as

Power of a test = $(1-\beta)$

where β is the probability of committing type II error.

Q.22. What do you know about the power function of a test ?

Answer

Let Ω_θ be the parametric space and $\Omega\theta_o$ be the sub-parameter space such that $\Omega_{\theta_0} \subset \Omega_\theta$. The probability 6 $(\theta) = P_\theta (C)$ regarded as a function of θ, $(\theta \in \Omega_\theta)$ is called the power function of the test and is given as

P_θ (c) = 1 - Probability of type II error associated with θ if $\theta \notin \Omega_{\theta_0}$

Q.23. Is it possible to minimise both the errors (type I and type II) simultaneously ? If not, how the power of a test is tried to increase ?

Answer

In a non-sequential sampling scheme, it is not possible to minimise the probability of both the error simultaneously. Any test which minimises the probability of error infact maximises the probability of error II. Hence during a test construction, one keeps the probability of error I fixed and tries to minimise the probabilities of error II i.e. minimises β and ultimately the power $(1-\beta)$ of the test is incresased.

Q.24. Does the power of a test depend upon the alternative hypothesis as well a on C. R. also ? What is the critieria for comparing two tests.

Answer

The power of a test is the function of alternative hypothesis and it also depends on the critical region. To compare the two tests, one by keeping the other things identical, has to compare the power of the two tests.

Q.25. What is the main job one has to do while dealing with the problem of testing of hypothesis ?

Answer

While dealing with the problem of testing of hypothesis, one has to mainly determine the size and location of critical regions.

Q.26. What do you mean by best critical region and most powerful test ?

Answer

If there are many critical regions (C. R's) having the same size α (say) for test then the C. R. with minimum β i.e. maximum power is termed as the best critical region and the test corresponding to this C. R. is termed as the mose powerful (M. P.) test.

Q.27. What is uniformly most powerful C.R. (U.M.P.C.R.) ?

Answer

A critical region C is the uniformly most powerful C.R. of size α for testing the null hypothesis $H_0 : \theta = \theta_0$ against an alternative composite hypothesis $H_1 : \theta \neq \theta_0$. if the set C is the best C.R. of size α for testing H_0 against each simple hypothesis in H_1.

Q.28. What do you mean by an uniformly most powerful test?

Answer

A test based on an uniformly most powerful critical region C of size α is termed as the uniformly most powerful test (UMP test) with significant level α for testing H_0 against a alternative composite hypothesis H_1.

Q.29. Does an uniformly most powerful test always exist ?

Answer

No, it does not always exist. However, if and when it exists, the Neyman Pearon Lemma provides a technique to find it.

Q.30. What do you know about an unbiased C.R. and an unbiased test ?

Answer

A critical region C of size α associated with a test of which the power does not fall below α, is said to be unbiased critical region and a test based on it is termed as unbiased test.

Q.31. When a test is said to be uniformly unbiased or completely unbiased ?

Answer

A test is said to be uniformly unbiased or completely unbiased if its power is always greater than the size of the C.R. A most powerful test is unbiased.

Q.32. Suppose that an UMP test does not exist, then what will be the alternative criteria in that situation ?

Answer

If in some cases, an UMP test fails to exist then in that situation one should adopt the critieria of an unbiased test as an alternative one.

Q.33. Define uniformly most powerful unbiased critical region.

Answer

Let C′ be any arbitrary unbiased critical region of size *a*. The region C* is said to be uniformly most powerful unbiased critical region for testing $H_0 : \theta = \theta_0$ against $H_1 : \theta \neq \theta$ if

$$P_{\theta 0}(c^*) = \alpha \quad \text{...............(a)}$$

$$P_{\theta}(c^*) \geq \alpha \ \forall\ \theta \neq \theta_0 \quad \text{...............(b)}$$

$$P_{\theta}(c^*) \geq P_{\theta}(c') \ \forall\ \theta \neq \theta_0 \quad \text{...............(c)}$$

where the arbitrary critical region C′ satisfies the relation A and B.

or

In other words if C is a set of unbiased critical regions of size *a* then any region $C^* \in C$ is said to be uniformly most powerful unbiased critical region for testing the hypothesis $H_0 : \theta = \theta_0$ against $H_1 : \theta \neq \theta$ if

$$\text{Power } (C^*) \geq \text{ Power } (C') \ \forall\ C' \in C$$

Q.34. In what four different ways, a statistical hypothesis is tested?

Answer

We test a statistical hypothesis in the following four different ways.

(1) Simple vs simple hypothesis

(2) Simple vs composite hypothesis

(3) Composite vs simple hypothesis

(4) Composite vs composite hypothesis

Q.35. How one can find the MPCR for testing the $H_0 : \theta = \theta_0$ vs $H_1 = \theta = \theta_1$.

Answer

Using Beyman Pearson Lemma, one can find the MPCR for testing $H_0 : \theta = \theta_0$ vs $H_1 : \theta = \theta_1$ where it is assumed that X_1, X_2,....., X_n are the continuous random variables with joint p d.f. f (x, θ)

Q.36. How one can get a UMPCR for testing $H_0\ \theta = \theta_0$ vs H_1 : $\theta \not\in \Omega\ \theta_1$ (composite hypothesis) ?

Answer

Using Neyman Pearson Lemma, one can get a uniformly most powerful critical region for testing $H_0 : t = t_0$ vs H_1 : ...... (composite alternative hypothesis).

Q.37. Does Neyman Pearson Lemma hold in case of discrete distributions (discrete random variables).

Answer

Yes, Neyman Pearson Lemma holds in case of discrete as well as contibuous distributions.

Q.38. Is there any basic difference between the methods of finding MPCR or UMPCR in cases of continous and discrete random variables ?

Answer

The method of finding the MPCR or UMPCR in case of continuous as well as discrete random variables are basically similar.

Q.39. Does a MPCR or UMPCR of exact size α always exist for testing $H_0 : \theta = \theta_0$ in case of discrete random variable ?

Answer

Such critical regions of exact size α may or may not exist for testing $H_0 : \theta = \theta_0$ in case of discrete random varibles.

Q.40. Let $x_1, x_2, \ldots\ldots, x_n$ be a random sample drawn from $N(\mu, \sigma^2)$ *where* μ is unknown but σ^2 is known one has to test $H_0 : \mu = \mu_0$ vs $H_1\ \mu = \mu_1 > \mu_0$. Then what will be the MPCR and UMPCR ?

Answer

We know that the sample mean $\bar{x} \sim N\left(\mu, \frac{\sigma^2}{n}\right)$. Let α be the size of CR then MPCR will be given as

$$W_0 = \left\{X/\bar{x} > \mu_0 + Z\alpha.\frac{\sigma}{\sqrt{n}}\right\} \qquad \ldots\ldots\ldots\ldots(1)$$

where $Z\alpha$ is the critical value of standard normal variate Z at α% level of significance.

Remark : We note that the critical region W_0 given by (1), of size α for testing $H_0 : \mu = \mu_0$ vs $H_1\ \mu = \mu_1 > \mu_0$ does not explicitly depend on μ_1 and hence the critical region W_0 obtained above is same for all $\mu_1 > \mu_0$. This implies that W_0 (1) is also the UMPCR for testing $H_0 : \mu = \mu_0$ against the composite (one sided) alternative $H_1 : \mu > \mu_0$.

Q.41. In case of the problem discussed in question number (40) what will be the MPCR and UMPCR for testing the hypothesis $H_0 : \mu = \mu_0$ vs $H_1 : \mu = \mu_1 < \mu_0$?

Answer

In this case, the MPCR, will be given as

$$W_0^1 = \left\{X/\bar{x} < \mu_0 - Z\alpha.\frac{\sigma}{\sqrt{n}}\right\} \qquad \ldots\ldots\ldots(2)$$

and likewise this MPCR W_0' (2) will also be the UMPCR for testing $H_0 : \mu = \mu_0$ vs the one sided composite alternative $H_1 : \mu = \mu_1 < \mu_0$.

Q.42. Does there exist any UMPCR for testing the hypothesis $H_0 : \mu = \mu_0$ against the composite (two sided) alternative $H_1 : \mu \neq \mu_0$ in a problem discussed in Q (40) ?

Answer

In case of two sided composite alternative hypothesis $H_1 : \mu \neq \mu_0$, there does not exist any UMPCR for testing $H_0 : \mu = \mu_0$. which is quite clear from the results obtained in questions (40) and (41).

In short for the two alternatives say $H_1: \mu > \mu_0$ and $H_1: \mu < \mu_0$, the two critical regions obtained are not the same

Q.43. Let $x_1, x_2, \ldots, x_n$ be a random sample of size n drawn from a population N (μ, σ^2) *where* μ is known but σ^2 is unknown. One has to test the hypothesis $H_0 : \sigma^2 = \sigma_0^2$ vs $H_1 : \sigma^2 = \sigma_1^2 > \sigma_0^2$ then what will be the MPCR and UMPCR?

Answer

Let α be the size of critical region then the MPCR is given by

$$W_0 = \left\{ X/, \sum_{i=1}^{n} \left(\frac{X_i - \mu}{\sigma_0} \right)^2 > X^2_{\alpha,n} \right\} \qquad \ldots\ldots\ldots(3)$$

It is noted that the MPCR given by (3) of size α for testing the $H_0 : \sigma^2 = \sigma_0^2$ vs $H_1 = \sigma^2 = \sigma_1^2 \, (\sigma_1^2 > \sigma_0^2)$ does not explicitly depend on σ_1^2 and so is the same for all such $\sigma_1^2 > \sigma_0^2$. Hence the C.R. (3) of size, is also UMPCR for $H_0 : \sigma^2 = \sigma_0^2$ vs $H_1 : \sigma^2 = \sigma_1^2 \; (\sigma_1^2 > \sigma_0^2)$, the composite (one sided).

Q.44. In case of problem discussed in (43), what will be the MPCR and UMPCR for testing $H_0 : \sigma^2 = \sigma_0^2$ vs' $H_1 : \sigma^2 = \sigma_1^2$ $(\sigma_1^2 < \sigma_0^2)$?

Answer

In this case the MPCR will be given as

$$W_0^1 = \left\{ X / \sum_1^n \left(\frac{x_i - \mu}{\sigma_0} \right)^2 < X^2_{1-\alpha, n} \right\} \qquad \text{.........(4)}$$

and likewise, this C.R. (4) will be also be UMPCR for testing $H_0 : \sigma^2 = \sigma_0^2$ vs composite alternative (one sided) $H_1 : \sigma^2 = \sigma_1^2$ $(\sigma_1^2 < \sigma_0^2)$.

Q.45. In a problem discussed in Q (43) does there exist any UMPCR for testing $H_0 : \sigma^2 = \sigma_0^2$ vs a two sided composite alternative $H_1 : \sigma^2 \neq \sigma_0^2$?

Answer

In case of two sided composite alternative $H_1 : \sigma^2 \neq \sigma_0^2$ there does not exist any UMPCR for testing $H_0 : \sigma^2 = \sigma_0^2$. which is quite clear from the C.R's obtained in (3) and (4).

It shows that for the two alternatives $H_1 : \sigma^2 > \sigma_\alpha^2$ and H_1' $\sigma^2 < \sigma_0^2$, the two critical regions (3) & (4) obtained are different and one thinks for type A critical region.

Q.46. Let there exists a MPCR and UMPCR for testing H_o: $\mu = \mu_o$ vs $H_1 : \mu = \mu_1$ $(\mu_1 > \mu_0$ or $\mu_1 < \mu_0)$ It is unnecessary for such MPCR and UMPCR to be unbiased also ?

Answer

Yes, such MPCR and UMPCR are necessarily unbiased for testing $H_0 : \mu = \mu_0$ vs $H_1 : \mu = \mu_1$ $(\mu_1 > \mu_0)$ or $(\mu_1 < \mu_0)$.

Q.47. For testing $H_0 : \mu = \mu_0$ against $H_1 : \mu \neq \mu_0$, there does not exist a UMPCR of size α then what will be the other type of critical region to be investigated ?

Answer

In case if the UMPCR fails to exist for testing $H_0 : \mu = \mu_0$ against $H_1 : \mu \neq \mu_0$, one should look for a uniformly most powerful unbiased critical region (UMPUCR) of the size α. This UMPUCR may be regarded as the best critical region available but it is not an easy task to construct such critical region.

Q.48. Consider the case of random sampling from $N(\mu, \sigma^2)$ *where* μ is unknown but $\sigma^2 > 0$ is known. What will be the C.R. for testing $H_0 : \mu = \mu_0$ against $H_1 : \mu \neq \mu_1$ of size α ?

Answer

Let xi (i = 1, 2, ..., n) be random sample of size *n* drawn from $N(\mu, \sigma^2)$. Then the critical region known as type A region is given as

$$W_0 = \left\{ X / \bar{x} < \mu_0 - Z_{\alpha/2} \frac{\sigma}{\sqrt{n}} \text{ or } \bar{x} > \mu_0 + Z_{\alpha/2} \frac{\sigma}{\sqrt{n}} \right\}$$

$$or\ W_0 = \left\{ X / \frac{|\bar{x} - \mu_0|}{\sigma/\sqrt{n}} > Z_{\alpha/2} \right\}$$

where $Z_{\alpha/2}$ is the critical value of Z at $\alpha/2$ % level of significance.

Q.49. Consider the case of random sampling from $N(\mu, \sigma^2)$ *where* μ is known but σ^2 is unknown. What will be the type A critical region for testing $H_0 : \sigma^2 = \sigma_0^2$ vs $H_1 : \sigma^2 \neq \sigma_0^2$ of size α ?

Answer

In such problem, the type A critical region of size α is given as

$$W_0 = \left\{ X / P\left(\sum_i^n \frac{(x_i - \mu)^2}{\sigma_0^2} < X_1^2, \text{ or } \sum_i^n \frac{(x_i - \mu)^2}{\sigma_0^2} > X_2^2 \right) = \alpha \right\}$$

Q.50. While dealing with the problem of hypothesis testing, what do you mean by a critical function ϕ (x).

Answer

The critical function ϕ (x) for a given x, represent the probability of rejecting the null hypothesis H_0 when the actual observations are x : $\{x_1, x_2, \ldots, x_n\}$.

Q.51. What do you mean by randomised test ?

Answer

While dealing with the problem of testing a hypothesis, if a set of observations x on $x_1, x_2, \ldots, x_n$ is obtained, one has to perform a random experiment which has two possible outcomes (say E_1 leading to the rejection of H_0 and E_2 leading to its acceptance) and in which E_1 has the probability ϕ (x) to occur. A test of this type is known as a randomised test.

Q.52. In case of $x_1, x_2, \ldots, x_n$ being continuous random variables, is there any similarity between the MP randomised test and MP non-randomised test for testing H_0 against H_1 ?

Answer

In such a situation, the MP randomised test will generally be identical with the MP non-randomised test. This is because in such a case $y = f_{\theta 1} (X) / f_{\theta 0} (X)$ too is likely to be a continuous random variable so that the probability of the equality $f_{\theta 1} (X) = K f_{\theta 0} (X)$will be zero.

Q.53. If $x_1, x_2, \ldots, x_n$ are discrete random variables and MPCR of size α for testing H_0 against H_1 exist, then is there any relation between MP randomised test and MP non randomised test ?

Answer

In such a situation also, the MP randomised test will be equivalent to the MP non-randomised test.

Q.54. Suppose it turns out that no C.R. of exact level α exist which is MP then in what way one should proceed in such a situation?

Answer

In such a situation, one may take instead a slightly lower level for which an MPCR does not exist and make use of the non-randomised test. But if in such a situation, one uses the randomised test then randomisation will be really worth while. For the MP randomised test will lead to an increase in power. While still effecting the desired level of control on the probability of error I.

Q.55. If a sufficient statistic T exists for the family $\{f(\theta), \theta \varepsilon \Omega\theta\}$ where $\Omega\theta$ is the parameter space then is there any relation between Neygman Pearson MP test and the sufficient statistic T ?

Answer

In such situation, Neyman Pearson MP test is essentially is a function of sufficient statistic.

More on Testing of Hypothesis

Q.56. Let the null hypothesis H_0 be a composite and concerned with parametric hypothesis then in what form it will be stated ?

Answer

A composite null hypothesis H_0 related to the parametric hypothesis will be stated in the form

$$H_0 : \theta \varepsilon \Omega_{\theta_0} \quad \text{.............(1)}$$

where Ω_{θ_0} is a subset of parameter space $\Omega_{\theta_0\,o}$ consisting more than a single point.

Q.57. What do you mean by a similar region concerning a problem of test of hypothesis ?

Answer

Let w be a C.R. of size α, (0 < α <1). Then we require that

$$P_\theta(w) = \alpha, \forall \theta \varepsilon \Omega_{\theta_0} \quad \text{................(2)}$$

i.e. a critical region w whose size is independent of $H_0 : \theta \varepsilon \Omega_{\theta_0}$ and is equal to the level of significance α, is called similar region for testing the null hypothesis $H_0 : \theta \varepsilon \Omega_{\theta_0}$

Q.58. In case of composite null hypothesis, what are the three essential stages for selecting a suitable test ?

Answer

The three essential stages for selecting a suitable test in such a case are as

(1) Finding all similar regions

(2) Finding those similar regions which are of size α.

(3) Selecting that similar region of size a among similar regions of size α that is the best from the point of view of power.

Q.59. Let $x_1, x_2, ..., x_n$ be a random sample from N (μ, σ^2) where μ and σ^2 both are unknown. Consider the problem of testing the composite null hypothesis $H_0 : \mu = \mu_0$ vs composite $H_1 : \mu = \mu_1 > \mu_0$. What will be the MP similar region of size a ?

Answer

Let we have the following critical region for the given test of hypothesis defined as

$$W_0 = \left\{ X / \frac{(x - \mu_0)}{s/\sqrt{n}} > t_{\alpha,(n-1)} \right\} \qquad ....(3)$$

which is independent of σ^2 and hence it is the MP similar region of size α for testing H_0 against H_1. Here tα (n-1) is the upper point of the t distribution with (n-1) d.f.

Q.60. Consider the situation discussed in question 4 and one has to test the composite null hypothesis $H_0 : \mu = \mu_0$ vs $H_1 : \mu = \mu_1 < \mu_0$. What will be the MP similar region ?

Answer

Let in this case, W_0^1 be the MP similar region then it will given as

$$W_0^1 = \left\{ X / \frac{(x - \mu_0)\sqrt{n}}{s} < -t_{\alpha(n-1)} \right\} \qquad ....(4)$$

Q.61. If in the problem discussed in question (4) one has to find the uniformity most powerful (UMP) similar region of size α, then what it will be ?

Answer

Since the MP similar region W_0 given in question 4, is independent of μ_1 i.e. is the same for all $\mu_1 > \mu_0$, it is infact. The UMP similar region of size α for testing $H_0 : \mu = \mu_0$ against the more general composite hypothesis $H_1 : \mu > \mu_0$.

Remark : In the similar fashion as discussed in question 6, the MP similar region Wo′ is also the UMP similar region of size α for testing $H_0 : \mu = \mu_0$ against the alternative $H_1 : \mu < \mu_0$

Q.62. Suppose $x_1, x_2, \ldots, x_n$ is a random sample from N (μ, σ^2). *where* both μ and σ^2 are unknown. Consider the problem of testing of the composite null hypothesis $H_0 : \sigma^2 = \sigma_0^2$ vs composite $H_1 : \sigma^2 = \sigma_1^2 > \sigma_0^2$.

What will be the MP similar region of size α for testing H_0 agaisnt H_1 ?

Answer

As we know that $\sum_{l=1}^{n} \frac{(x_i - \overline{x})^2}{\sigma_0^2} \sim \chi^2_{(n-1)}$

we define W_o as CR gieven as

$$W_0 = \left\{ X / \sum_{i}^{n} \frac{(x_i - \overline{x})^2}{\sigma_o^2} > \chi^2_{\alpha,(n-1)} \right\} \qquad \text{..........(5)}$$

where $\chi^2_{\alpha,(n-1)}$ is the upper critical value of χ^2 distribution at $\alpha\%$ significant level with (n-1) d.f. This W_0 is independent of μ, it is the MP similar region of size α for testing H_0 against H_1.

Q.63. If in the case discussed in question number 7, one has to test the composite null hypothesis $H_0 : \sigma^2 = \sigma_0^2$ vs composite $H_1 : \sigma^2 = \sigma_1^2 < \sigma_0^2$. what will be the MP similar region of size α ?

Answer

In this problem, the MP similar region of size α for testng H_0 against H_1 will be given as

$$W_0^1 = \left\{ X / \sum_{i}^{n} \frac{(x_i - \bar{x})^2}{\sigma_0^2} < X^2_{\alpha,(n-1)} \right\} \qquad \text{..........(6)}$$

where $X^2_{\alpha,(n+1)}$ is the lower α point of χ^2 distribution for (n-1) d.f.

Q.64. For the problem discussed in question number 7, what will be the UMP similar region of size α for testing H_0 against H_1?

Answer

Since the MP similar region W_0 given in (7) is dependent of all $\sigma_1^2 > \sigma_0^2$, it is infact UMP similar region of size a, for testing $H_0 : \sigma^2 = \sigma_0^2$ against one sided alternative $H_1 : \sigma_1^2 > \sigma_0^2$

Q.65. Like the question no. (9), what will be the UMP similar region of size α for testing the $H_0 : \sigma^2 = \sigma_0^0$ vs $H_1 : \sigma_1^2 < \sigma_0^2$?

Answer

Since the MP similar region W_0^1 given in (6) is independent of all $\sigma_1^2 < \sigma_0^2$ this W_0^1 is also the UMP similar region for testing H_0 against H_1.

Q.66. What do you mean by the monotonicity property of a power function ?

Answer

Suppose one has to test the hypothesis $H_0 : \theta = \theta_0$ vs. $\theta \neq \theta_0$ and let W be the chosen CR of size α. If the power function, say P_θ (W) increases monotonically as θ moves further and further away from the θ_0 then one can says that the power function P_θ (W) follows the monotonicity property.

Q.67. Can we say that all the UMP tests and UMPU tests if they exist, follow the monotonicity property ?

Answer

Yes, all the UMP tests as well as UMPU tests if they exist, have the monotonicity property.

Q.68. What do you mean by the consistency property of a power function ?

Answer

Let us denote by Wn, the CR for testing the hypothesis $H_0 : \theta \in \Omega_{\theta_0}$ against $H_1 : \theta \in (\Omega_\theta - \Omega_{\theta_0})$. A sequence of tests corresponding to the sequence of CR (Wn) is said to be consistent if power $P_\theta (W_n) \to 1$ as $n \to \infty$ $\forall \theta \hat{I} (\Omega_\theta - \Omega_{\theta_0})$.

Q.69. Do all the UMP tests and UMPU tests follow the consistency property ?

Answer

In fact all the UMP tests as well as UMPU tests for testing H_0: $\theta = \theta_0$ are consistent, which is obvious from the following theorem.

Theorem : A consistent estimator t_n for a parameter θ yeilds a consistent sequence of tests for testing the hypothesis $H_0 : \theta = \theta_0$.

Q.70. What do you know about likelihood ratio test ?

Answer

Suppose one has to test a simple hypothesis $H_0 : \theta = \theta_0$ against a composite alternative $H_1 : \theta \neq \theta_0$. Then the test based on the CR W = $\{x/\lambda (x_1, x_2,...., x_n) < \lambda_0\}$ is called the likelihood ratio test for L being the likelihood ratio defined as

$$\lambda(x_1, x_2..........x_n) = \frac{L(\theta_0/x_1, x_2..........x_n)}{\text{Sup}\,\theta \in \Omega_\theta\, L(\theta/x_1, x_2..........x_n)}$$

and λ_0 is a constant chosen such a way that the probability of first kind of error is equal to α.

Q.71. Let $x_1, x_2,...., x_n$ be a random sample from N (μ, σ^2) where μ is unknown but σ^2 is known. Then what would be the likelihood ratio test for testing $H_0 : \mu = \mu_0$ vs $H_1 : \mu \neq \mu_0$?

Answer

In such a problem of testing the hypothesis H_0, the likelihood ratio test will be given by the critical region W defined as

$$W = \left\{ x / \left| \frac{(\overline{x} - \mu_0)\sqrt{n}}{\sigma} \right| > Z_{\alpha/2} \right\}$$

where $Z_{\alpha/2}$ is the upper $\alpha/2$ point of the standard normal distribution.

Q.72. For the problem discussed in question 16, what would be the likelihood ratio test for testing the $H_0 : \mu = \mu_0$ vs $H_1 : \mu > \mu_0$

Answer

In this problem, the alternative hypothesis is one sided (right side) composite one. In such problem, the likelihood ratio test will be given by the critical region W_0 defined as

$$W_0 = \left\{ x / \frac{(x - \mu_0)\sqrt{n}}{\sigma} > Z_\alpha \right\}$$

where $Z\alpha$ is the upper α point of Z distribution.

Q.73. If someone wants to test the hypothesis $H_0 : \mu = \mu_0$ against $H_1 : \mu < \mu_0$ for the problem discussed in question 16, what would be the likelihood ratio test for that ?

Answer

Here the alternative hypothesis is one sided (left) Composite one and hence for likewise, the critical region W_0^1, the likelihood ratio test will be given by critical region W_0^1 defined as

$$W_0^1 = \left\{ x / \frac{(x - \mu_0)\sqrt{n}}{\sigma} < -Z_\alpha \right\}$$

where $-Z\alpha$ is lower α point of Z distribution.

Q.74. If $x \sim N(\mu, \sigma^2)$ where μ is unknown but σ^2 is known and one has to test the hypothesis $H_0 : \mu = \mu_0$ against one sided alternative hypothesis either $H_1 : \mu > \mu_0$ or $H_1 : \mu < \mu_0$. Then, is there any relation between the likelihood ratio tests and the corresponding UMP tests ?

Answer

In such a problem, the likelihood relation and the corresponding UMP tests are identical.

Q.75. Is there any relation between the likelihood ratio test and the corresponding UMPU test for testing the hypothesis $H_0 : \mu = \mu_0$ against the two sided alternative hypothesis $H_1 : \mu \neq \mu_0$?

Answer

In such a problem, the both types of tests are identical.

Q.76. Let $x_1, x_2, \ldots, x_n$ be a random sample from N (μ, σ^2) where μ is known but σ^2 is unknown. What will be the LR test for testing $H_0 : \sigma^2 = \sigma_0^2$ vs $H_1 : \sigma^2 = \sigma_1^2 > \sigma_0^2$.

Answer

Let α be the size of the critical region. Then the LR test will be given by the critical region W_o defined as

$$W_0 = \left\{ x / \sum_{i}^{n} \left(\frac{x_i - \mu}{\sigma_0} \right)^2 > \chi^2_{\alpha,n} \right\}$$

where $\chi^2_{\alpha,n}$ n is the upper α point of χ^2 distribution with d.f. equal to n, the size of sample itself.

Q.77. Consider the situation discussed in question 21, what will be the LR test criterion for testing $H_0 : \sigma^2 = \sigma_0^2$ vs $H_1 : \sigma^2 = \sigma_1^2 < \sigma_0^2$?

Answer

For testing such type of hypothesis, the LR test criteria will be given by a CR W_0^1 defined as

$$W_0^1 = \left\{ x / \sum_{l=1}^{n} \left(\frac{(x_i - \mu}{\sigma_0} \right)^2 < \chi^2_{1-\alpha,n} \right\}$$

where $\chi^2_{1-\alpha,n}$ is the lower (1-α) point of χ^2 distribution with d.f. equal to n.

Q.78. Suppose one has to test the hypothesis $H_0 : \sigma^2 = \sigma_0^2$ vs H_1: $\sigma^2 = \sigma_1^2 > \sigma_0^2$ in the situation as discussed in question 21. Then, is there any relation among the LR criteria, MPCR and UMPCR ?

Answer

In this situation, the LR criteria is identical to MPCR and UMPCR both.

Q.79. Suppose one has to test $H_0 : \sigma^2 = \sigma_0^2$ vs $H_1 : \sigma^2 = \sigma_1^2 < \sigma_0^2$ in the situation discussed in question 21. What will be the relation among LR test criteria, MPCR and UMPCR ?

Answer

Here again, like in question 23, LR test criterion will be identical to critical regions MPCR and UMPCR.

Q.80. If in question 21, one has to test the hypothesis $H_0 : \sigma^2 = \sigma_0^2$ vs $H_1 : \sigma^2 \neq \sigma_0^2$ then what will be the LR test criteria ?

Answer

In such a problem of testing of hypothesis, the LR test criteria is given by

$$W^* = \left\{ X / P\left[\sum_{l=1}^{n} \left(\frac{X_i - \mu_0}{\sigma_0} \right)^2 < \chi_1^2 + \sum_{l=1}^{n} \left(\frac{X_i - \mu}{\sigma_0} \right)^2 > \chi_2^2 \right] = \alpha \right\}$$

which is identical with type A critical region.

Q.81. If the null hypothesis H_0 is itself a composite hypothesis say $H_0 : \theta \in \Omega_{\theta_0}$ then how the likelihood ratio λ is defined ?

Answer

If null hypothesis H_0 is itself a composite hypothesis say $H_0 : \theta \in \Omega_{\theta_0}$ alternative $H_1 : \theta \in \Omega_{(\theta - \theta_0)}$ then the likelihood ratio λ is defined as

$$\lambda(x_1, x_2, ...x_n) = \frac{\text{Sup } \theta \in \Omega\theta_0 \; L(\theta/x_1, x_2......., x_n)}{\text{Sup } \theta \in \Omega\theta \;\; L(\theta/x_1, x_2......., x_n)}$$

and the CR of the test is given as

$$W = \{X/\lambda(x_1, x_2, ..., x_n) < \lambda_0\}$$

the constant λ_0 being so chosen that size of the CR may be equal to α.

Q.82. Let $x_1, x_2, ..., x_n$ be a random sample from $N(\mu, \sigma^2)$ where both μ and σ^2 are unknown. Consider the problem of testing the null hypothesis $H_0 : \mu = \mu_0$ vs $H_1 : \mu > \mu_0$. What will be the CR of the LR test?

Answer

For testing the null hypothesis $H_0 : \mu = \mu_0$ against one sided (right) composite alternative hypothesis $H_1 : \mu > \mu_0$, the CR of LR test will be given as

$$W_1 = \left\{ x / \frac{(\overline{x} - \mu_0)\sqrt{n}}{s} > t_{\alpha, n-1} \right\}, s^2 = \frac{1}{n-1}\sum_{1}^{n}(x_i - \overline{x})^2$$

Q.83. Consider the situation discussed in question 27, for testing the null hypothesis

$H_0 : \mu = \mu_0$ vs $H_1 : \mu < \mu_0$

Then what will be the CR of the LR test ?

Answer

For testing such hypothesis, against a one sided (left) composite alternative hypothesis, the CR of the LR test will be given as

$$W_2 = \left\{ x / \frac{(\overline{x} - \mu_0)\sqrt{n}}{s} < -t_{\alpha, n-1} \right\}$$

Q.84. Consider the situation discussed in question 27 and if one has to test the null hypothesis $H_0 : \mu = \mu_0$ vs $H_1 : \mu \neq \mu_0$ then what will be the CR of the LR test ?

Answer

If in this situation, one has to test the null hypothesis $H_0 : \mu = \mu_0$ against two sided composite alternative hypothesis $H_1 : \mu \neq \mu_0$ then CR of the corresponding LR test will be given as

$$W = \left\{ X / \left| \frac{(x - \mu_0)\sqrt{n}}{s} \right| > t_{\alpha/2, n-1} \right\}$$

where t $\alpha/2$ is the upper $\alpha/2$ point of t distribution with (n-1) d.f.

Q.85. Is there exists any relation between the student's t test and LR test ?

Answer

The critical regions given for the problem of test of hypothesis discussed in questions 27, 28 and 29 show that the student's t tests are the likelihood ratio tests.

Q.86. Let there be K random samples from K normal distributions with means μi (i = 1, 2, ..., K). Further let these random samples be independent of each other. Consider the problem of testing the hypothesis $H_0 : \mu_1 = \mu_2 = \mu_3 = ... = \mu_k$

against the H_1 : μ's are not equal. Assume that variances of the distributions which are supposed to be unknown, are all equal. Then what will be the LR test ?

Answer

Let SB be the sum of square between samples with d.f. equal to (k-1) and S_w be the sum of square within the samples with d.f equal to (n-K). Then under H_0, the LR test is given by F statistics defined as

$$F = \frac{S_{B/(k-1)}}{S_{w/(n-k)}} \sim F_{(k-1,n-k)} = \frac{MSB}{MSW}$$

and the CR for LR test is given by

$$W = \left\{ X / \frac{MSB}{MSW} > F_{\alpha,(k-1,n-k)} \right\}$$

where $F_{\alpha,\ (k-1,\ n-k)}$ is the upper α point of F distribution with (k-1, n-k) d.f and is known as ANOVA test.

Q.87. Does the LR test follow the assymptotic property ?

Answer

Yes, Likewise the maximum likelihood estimator for the parameter θ, the likelihood ratio test also follows the assymptotic property.

Q.88. Does the LR test follow the consistency property ?

Answer

Yes, likewise the MLE, for the parameter θ, the LR test also follows the consistency property.

Q.89. What do you mean by a CR of type A ?

Answer

Neyman and Pearson called a CR corresponding to a UMPU test as a CR of type A.

Q.90. What is the advantage of UMPU test over UMP test ?

Answer

The advantage of UMPU test over UMP test is that even when UMP tests do not exist, the UMPU test may exist.

Q.91. In what sense, an UMPU test is said to be admissible ?

Answer

An UMPU test is said to be admissible in the sense that there can not exist another test which is atleast as powerful as the UMPU test against all alternatives and more powerful against some one.

Q.92. What do you know about the uniqueness property of BCR determined by Neyman Pearson principle ?

Answer

If both H_0 and H_1 are simple then the corresponding region determied by Neyman Pearson principle is unique provided it is determined by minimising β subject to the condition that α is fixed.

Q.93. If both H_0 and H_1 are simple then can the LR method give the corresponding BCR ?

Answer

Yes, if both H_0 and H_1 are simple, then the LR method can always give the corresponding best critical region.

Q94. If the null hypothesis H_0 is simple but the alternative H_1 is composite and if among the standard BCR, there exists a UMPCR, then can it be determined by LR principle ?

Answer

Yes, in such a situation, if there exists a UMPCR, then LR principle will determine that region.

Q.95. If an UMPCR exists for a test what is your opinion for the null hypothesis to be either simple or composite ?

Answer

In an UMPCR, the H_0 may or may not be simple.

Q.96. What type of relation exists between the power of a BCR and its size α, for testing α simple H_0 against simple H_1 ?

Answer

The power of a BCR is always greater or equal to its size α for testing a simple H_0 against simple H_1.

Q.97. What do you mean by a uniformly most powerful unbiased (UMPCR) test ?

Answer

A uniformly most powerful unbiased test is one which is as powerful or more powerful than any other unbiased test of equal size with respect to all alternative hypothesis.

Q.98. If some one accepts the principle of unbiasedness then what type of test will be most advantageous ?

Answer

If the principle of unbiasedness is accepted and an UMPU test exists, then it is obvious that this UMPU test is the most advantageous to use.

Q.99. There is a test procedure which is closely akin to the maximum likelihood estimation procedure. What is that test procefure?

Answer

A test procedure which is closely akin (alike) to the maximum likelihood estimation procedure, is likelihood ratio test procedure reported by Neyman and Pearson.

Q.100. In a LR test procedure, if λ (x) is not considerably less than 1 then what type of decision about the null hypothesis is appropriate ?

Answer

In such a situation, the appropriate decision, is to accept the null hypothesis while for λ (x) being very smaller than unity, the rejection of null hypothesis is appropriate.

Q.101. Let $U = f(\lambda)$ be any monotonic function of λ in case of LR test. Can λ (x) be also used for testing purpose ?

Answer

Yes, The test based on U is equivalent to the likelihood ratio test. The CR for the test is based on $f(0) < U < f(\lambda_o)$, $f(\lambda 0) < U < f(0)$.

Q.102. Consider a random sampling from the rectangular distribution

$F_\theta(x)=\left\{\begin{smallmatrix}1\\ 0\end{smallmatrix} \text{otherwise} \quad \text{if} \quad \theta \le x \le \theta+1\right\}$

If one has to test the hypothesis $H_0:\theta=\theta_0$ VS $H_1:\theta\neq\theta_0$ then does there exist a UMP for all alternatives ?

Answer

No, There exists a pair of UMP tests for one sided alternatives (i.e. $H_1 : \theta > \theta_0$ & $H_1 : \theta < \theta_0$) and hence no UMP test exists for all alternatives. (i.e. for two sided alternatives).

Q.103. Consider a random sampling from an exponential distribution with p d.f.

$f_\theta(x) = \left\{\begin{smallmatrix}\theta.e^{-\theta x}\\ 0 \quad \text{otherwise}\end{smallmatrix} \text{ if } x>0\,,\, 0<\theta<\infty\right\}$

The problem is to test the

(1) $H_0 : \theta = \theta_0$ vs $H_1 : \theta = \theta_1 > \theta_0$

(2) $H_0 : \theta = \theta_0$ vs $H_1 : \theta = \theta_1 < \theta_0$

(3) $H_0 : \theta = \theta_0$ vs $H_1 : \theta = \theta_1 \neq \theta_0$

Does there exist a UMP test for all the three tests ?

Answer

Actually , for one sided alternative expressed in (1) and (2), there exists a pair of UMP tests but for two sided alternatives there does not exist a UMP test.

Q.104. Let $x_1, x_2, \ldots, x_n$ be a random sample from a Poisson distribution with unknown parameter λ ($0 < \lambda < \infty$). Suppose α is such that a MP (non randomised) test for $H_0 : \lambda = \lambda_0$ vs $H_1 : \lambda = \lambda_1$ exists. What will be the distribution of test statistic which is used for defining the test ?

Answer

The test statistic used for defining such test will itself follow the Poisson distribution.

Q.105. Consider the problem discussed in question 49. Does an UMP test exist for testing $H_0 : \lambda = \lambda_0$ against one sided alternative ?

Answer

In case of one sided alternative either right or left sided, there always exists an UMP test for testing $H_0 : \lambda = \lambda_0$.

Q.106. Let $x_1, x_2, \ldots, x_n$ be a random sample from a N (μ, σ^2) with μ and σ^2 both unknown. Do the UMP similar tests against one sided alternatives and the UMPU similar test for $H_0 : \mu = \mu_0$ posses the monotonic power function and are consistent ?

Answer

Yes, in such a case, the UMP similar tests against one sided alternatives and the UMPU similar test for $H_0 : \mu = \mu_0$ posses the monotonic power functions as well as the consistency property.

Q.107. Consider the problem discussed in question (51), and if some one has to test the hypothesis about the parameter σ^2. Do the UMP similar tests against one sided alternatives and the UMPU similar test for $H_0 : \sigma^2 = \sigma_0^2$ have monotonic power functions and are consistent ?

Answer

Yes, for testing such hypothesis about the population parameter σ^2, the UMP similar tests against one sided alternatives and the UMPU similar test for $H_0 : \sigma^2 = \sigma_0^2$ posses the monotonic power functions and are consistenct.

Q.108. Two independent random samples are taken from two normal populations, say N (μ_1, σ_1^2) and N (μ_2, σ_2^2) , σ_1^2 and σ_2^2 being known. Let some one has to test the following hypothesis

(1) $H_0 : \mu_1 = \mu_2$ vs $H_1 : \mu_1 > \mu_2$

(2) $H_0 : \mu_1 = \mu_2$ vs $H_1 : \mu_1 < \mu_2$

(3) $H_0 : \mu_1 = \mu_2$ vs $H_1 : \mu_1 \neq \mu_2$

What is your idea for the UMP similar tests for the hypothesis in cases (1) and (2) and the UMPU similar test in case (3) to have monotonic power functions as well as the consistency property ?

Answer

The two UMP similar tests in case (1) and (2) and the UMPU similar test in case (3) will have certainly the monotonic power functions as well as the consistency property.

Q.109. Suppose that σ_1^2 and σ_2^2 are both unknown but are equal in the question (53). Can we say again that the two UMU similar tests and the UMPU similar test have consistency property with the monotonicity of their power funtions ?

Answer

Yes, in such a case also, all three tests have consistency property with the monotonocity of their power functions.

Q.110. Consider the problem discussed in question 53 and let μ_1 and μ_2 be both known. Consider the following problems of test of hypothesis.

(1) $H_0: \sigma_1^2 = \sigma_2^2$ vs $H_1: \sigma_1^2 > \sigma_2^2$

(2) $H_0: \sigma_1^2 = \sigma_2^2$ vs $H_1: \sigma_1^2 < \sigma_2^2$

(3) $H_0: \sigma_1^2 = \sigma_2^2$ VS $H_1: \sigma_1^2 \neq \sigma_2^2$

What is your idea for the UMP similar tests for the hypothesis in the cases (1) and (2), UMPU similar test in the case (3) to have monotonic power function with consistency property?

Answer

Both the UMP similar tests and UMPU similar test will posses the monotonic power functions with the consistency property.

Q.111. In the problem discussed in question (53), Assume that μ_1 and μ_2 are unknown and all three hypothesis discussed in question (55) are considered. What property of the power function ?

Answer

In such a situation also, all the three power functions corresponding to similar tests will have monotonicity and consistency property.

Q.112. Consider n Bernoullian trials with probability of success p for each trial and LR test is derived for testing the following hypothesis.

$H_0 : p = p_0$ vs $H_1 : p > p_0$

$H_0 : p = p_0$ vs $H_1 : p < p_0$

$H_0 : p = p_0$ vs $H_1 : p \neq p_0$

Let $p_0 = \frac{1}{2}$, then can we say that the LR tests derived will be identical with the corresponding UMP and UMPU tests?

Answer

Yes, the LR tests derived in the three cases, will be identical with the corresponding UMP and UMPU tests for $p_0 = \frac{1}{2}$

Q.113. Let $x_1, x_2, \ldots, x_n$ be a random sample from a Poisson distribution with parameter λ and the LR tests are derived for testing the following hypothesis

(1) $H_0 : \lambda = \lambda_0$ vs $H_1 : \lambda > \lambda_0$

(2) $H_0 : \lambda = \lambda_0$ vs $H_1 : \lambda < \lambda_0$

(3) $H_0 : \lambda = \lambda_0$ vs $H_1 : \lambda \neq \lambda_0$

Do the three LR tests derived will be identical with the corresponding UMP and UMPU tests ?

Answer

Yes, the LR tests derived in all the three cases, will be identical with the corresponding UMP and UMPU tests.

Q.114. *Theorem* : If B (θ) is strictly monotonic function of a real parameter θ and $R(\theta)e^{B(\theta).N(x)}\,1(x)$ is the likelihood function of $x_1, x_2, \ldots, x_n$. To test the hypothesis.

$$H_0 : \theta \le \theta_0 \quad \text{vs } H_1 : \theta > \theta_0$$

or

$$H_0 : \theta \ge \theta_0 \quad \text{vs } H_1 : \theta < \theta_0$$

there exists a UMP test.

Q.115. Consider two normal populations $N(\mu_i, \sigma_i^2, i = 1, 2)$ Let n_1 and n_2 be the sizes of two independent samples drawn from two populations and assume that $n_1 \ne n_2$. Consider the following test of hypothesis.

$$H_0 : \sigma_1^2 = \sigma_2^2 \quad \text{VS } H_1 : \sigma_1^2 \ne \sigma_2^2$$

What is your opinion about the critical region with its unbiasedness ?

Answer

For testing this hypothesis, the F tests with CR as the union of two extreme tails of equal area is biased one.

Chapter **15**

Non Parametric Method of Estimation

Q.1. What is the most fundamental assumption to be followed by random variable for using the standard method of parametric statistics or parametric inference ?

Answer

The most fundamental assumption required to be followed by random varibale for using the parametric statistics to have statistical inference drawn from certain set of data is that the random variables have normal distribution.

Q.2. For what type of statistical inference to be drawn from the data, the parametric techniques are useful ?

Answer

The standard parametric techniques are useful for the estimation of parameters and for testing hypothesis concerning them.

Q.3. What is parametric statistic ?

Answer

Any sample statistic which is used for the estimation of a parameter and for testing hypothesis concerning it, for a population, is known as parametric statistic.

Q.4. If a considerable amount of collected data is such that the underlying distribution is not easily specified then what type of statistic (or statistical technique) is used to handle such data?

Answer

In such type of data where the underlying distribution is not easily specified, the distribution free statistic (non parametric method or technique) is appropriate in handling that data.

Q.5. Why a non parametric method is known as distribution free method?

Answer

The sampling distribution of a sample statistic which is used as non parametric statistic, does not depend on the distribution function of the population from which the sample is drawn and hence it is known as distribution free method.

Q.6. Does the sampling distribution of a non parametric statistic depend upon the parameters or parametric functions of the parent population from which the sample is drawn.

Answer

The sampling distribution of a non parametric statistic never depends upon the parameters of the parent population. Samping distribution is free from the parent parameter of the population.

Q.7. What is basic difference between the non parametric statistic and parametric statistic ?

Answer

The non parametric statistic compares distribution rather than parameters where as the parametric statistics compare parameters of a distribution rather than the distribution itself.

Q.8. To which characteristics, of the population, a non parametric statistic is more sensitive ?

Answer

These statistics may be sensitive to changes in location, in spread or in both characteristics of the population.

Remark : In general, we do not try to maintain a distinction between distribution free and non parametric statistics, but rather we call them all as non parametric statistics.

Q.9. When it is possible to make only weak assumptions about the nature of the distributions for underlying data, then out of the two type of statistics parametric and non parametric, which one is appropriate ?

Answer

In such situations, the non parametirc statistic is more appropriate than parametirc statistic.

Remark : Non parametric statistics are intended to apply to a large class of distribution rather than a single distribution or all possible distributions.

Q.10. If it is possible to do little more than categorize data because of lack of an adequate scale of measurement then what type of test is appropriate ?

Answer

In such a stituation, the non parametric test may be the best we can do.

Remark : In some cases of studies based on data, the characterization of data may be a way of collecting a lot of data in a hurry so that for such data the power of a non parametric test will be adequate for the needs of the investigator.

Q.11. When there is possibility to rank a available data, then what procedure will be appropriate ?

Answer

In such cases of ranked data, the non parametric procedures are available to handle them for scientific purposes. For example, food product may be ranked for texture or flavour, plants may be scored for virus infection or plots for insect infestation in a variety trial involving many locations, variance may be hetreogeneous in violation of the usual assumptions for a valid ANOVA and ranks may be the best measures for an analysis.

Q.12. If the observations are in the form of continuous, ranks, or the signs of differences for paired observations, then in that cases which type of statistics is appropriate and why?

Answer

In such cases of observations, the non parametric statistics are appropriate often, though not always because these are quick and easily to apply and to learn by the research scientists of various fields.

Q.13. In what kind of data or population followed by the data, the use of non parametric procedures is disadvantegous ?

Answer

If the form of the apparent population is known to be reasonably close to a distribution for which there is a standard theory or if the data can be transformed so that such is the case, then non parametric procedures extract less information than that is available in the data.

Q.14. If all the experiments of an investigator result in the data such that the H_0 is true then out of prametric and non-parametric tests, which one is better than other ?

Answer

In such cases, the non parameteric procedures are as good as parametric procedures, since the investigator sets the error rate.

Q.15. If the null hypothesis is false and the assumptions about the parent population are valid then out of the two procedures, which one is found to be better than other ?

Answer

In such a situation, the usual problem is to detect differences among means and classical procedures are found to be better than non parametric procedures.

Q.16. In case of small samples, say $n \leq 10$ which one out of two procedures, has relatively high efficiency ?

Answer

In such a situation, the efficiency of non parametric procedure is quite high as relatively to that of parametric procedures and it decreases as *n* increases. On the other hand, efficiency may not be important for very large *n*.

Q.17. In case of applying sign test, one considers median rather than mean then what is the test statistic used for ?

Answer

The sign test is based on the signs of differences between the paired values and the test statistic used is χ^2 defined as

$$\chi_1^2 = (n_1 - n_2)^2 \; / \; (n_1 + n_2)$$

where n_1 and n_2 are the number of + and - signs.

Q.18. What is the adjustment made for maintaining the continuity property of χ^2 distribution in case of using sign test statistic given as

$$\chi_1^2 = (n_1 - n_2)^2 \; / \; (n_1 + n_2)$$

Answer

For the required adjustment made in the χ^2 statistic one considers numerator $(|n_1 - n_2| - 1)^2$ in place of $(n_1 - n_2)^2$.

Q.19. In case of using sign test, one does not need homogeniety of variances in usual sense. Why it is so ?

Answer

In this case, the homogeniety of variances is not needed because each differences (Xi - Yi), i = 1, 2, ..., n may be from different continuous population provided that all the populations have zero as median.

Q.20. In what sense, a lot of information contained in the data is not utilized by sign test ?

Answer

In case of applying sign test, only the number of plus and minus signs of differences (Xi - Yi) of the paired observations (Xi - Yi), i = 1, 2, ...,n is considered regardless of the magnitude of the differences which is the demetric of the sign test.

Q.21. Can we detect a departure from the null hypothesis in case the number of pairs of observations being less than six.

Answer

In case of number of pairs of observations being fewer than six, it is not possible to detect the departure from null hypothesis.

Q.22. In case of a single population, if one has to test the null hypothesis that the median is a specified value say Md, then what test procedure will be adopted and also give the test statistic ?

Answer

In such case, one must use the sign test and the test statistic to be used is given as

$$X_1^2 = \frac{(n_1 - n_2)^2}{(n_1 + n_2)}$$

where n_1 and n_2 are the number of observatons in the sample which lies respectively above and below the specified value.

Q.23. Suppose for paired observations, one wants to test whether the treatment A gives a response that is C unit better than that given by the treatment B. Which test will be appropriate?

Answer

In such case, one should observe the sign of difference $(Y_{1i} - (Y_{2i} + C))$ and then must apply sign test. Here Y_{1i} and Y_{2i} are the response due to treatment A and B respectively.

Q.24. In what sense signed rank test is considered as an improvement on the sign test ?

Answer

A signed rank test is considered as an improvement on the sign test because it uses the magnitude of differences also in addition to the signs of differences.

Q.25. What is the difference between test criterion followed by signed rank test (NP test) and the other test based on Z, t, X^2 and F statistic ?

Answer

In case of signed rank test, the sum of negative or $+_{ve}$ ranks considered as test statistics being less than the tabulated value at α% level of significance, gives an evidence of significance of null hypothesis while in case of other tests like Z, t, X^2 and F, the calculated value of test statistic being less than the tabulated value, gives an evidence of non-significant of null hypothesis.

Q.26. Let T^+ and T^-, be the sum of positive and negative ranks in case of adopting signed rank test. Out of T^+ and T, which one will be taken as test statistic ?

Answer

Out of T^+ and T^-, the one having numerically the lower value (i.e. the value in magnitude) will be considered a test criteria or test statistic.

Q.27. If the number of pair of observations is beyond range of table (say n > 25 pairs), what will be the alternative procedure to be adopted in place of signed rank test ?

Answer

In such situation, one should adopt Z test based on fundamental area property which is given as

$$Z = (T - \mu_T)/\sigma_T$$

where

$$\mu_T = \frac{n(n+1)}{4}, \; \sigma_T = \sqrt{\frac{n(n+1)(2n+1)}{24}}$$

for n equal to the number of pairs

Q.28. What is the test criterion to be followed to reject the null hypothesis in case Wilcoxon Mann Whitney two sample test?

Answer

If the calculated value of test statistic (say T_1) is less than the table value at α % level sign. Then one should reject the null hypothesis.

Bibliography

1. Cochran, W.G. and Cos, G. M. (1957) : Experimental Designs, II, Edition, John Wiley and Sons, New York.
2. Fisher, R. A. (1947) : The Design of Experiment, Alive and Beyond, London.
3. Federer, W. T. (1967) : Experimental Design. The Macmillan Co. New York.
4. Goon, A. M. , Gupta, M. K. and Dasgupta, B. (1972) : An Outline of Statistical Theory, Vol II, The World Press Pvt. Ltd. Calcutta.
5. Gupta, H. C. and Kapoor, V. K. (1990) : Fundamental of Mathematical Statistics, Sultan Chand and Sons, New Delhi.
6. Gupta, H. C. and Kapoor, V. K. (1990) : Fundamental of Applied Statistics, Sultan Chand and Sons, New Delhi.
7. Kendal, M. G. and Stuart, A. (1967) : Advance Theory of Statistics, Inference and Relationship, Vol II, Charles, Griffin and Co. Ltd. London.
8. Murray, R. Spiegel (1961) : Theory and Problems of Statistics, Schaum Publishing Company, U. S. A.
9. Mood, A. M., Grajybill, F. A. and Bose, D. C. (1974) : Introduction to The Theory of Statistics, McGraw Hill. Kogakusha, Ltd. London.
10. Panse, V. G. and Sukhatme, P. V. (1967) : Statistic Methods for Agricultural Workers, I. C. A. R. , New Delhi.

11. Robert, G. D. Steel and Jones, M. Torrie (1980) : Principles and Procedures of Statistics (Second Ed.), McGraw Hill. Kogakusha, Ltd. London.

12. Saxena, H. C. and Surendran, P. U. (1990) : Statistical Inference S. Chand and Company Ltd. New Delhi.

13. Snedekor, G. W. and Cochran, W. G. (1989) : Statistical Methods, Iowa State University Press, Ames, Iowa, USA.

About the Authors

Dr. K.S. Kushwaha: He was an intelligent student and secured first rank in the "District Board Exam" (8th Class) in 1969 and achieved 5th rank on the basis of state level merit of U.P. State prepared during 1969. He was awarded a fellowship Rs.100/- month during High School and Rs.150/- month during Intermediate classes of his study. He was asked by the Honourable Governor, Dr. M. Chennareddy to take admission in Govt.'s Queen's Intermediate College, Varanasi. He passed High School Exams with honour and obtained B.Sc. degree (1975) and M.Sc. (Statistics) in 1977 from Science College, B.H.U., Varanasi in first division.

Dr. Kushwaha completed one year "Professional Statistician Certificate Course" (P.S.C.C.) from Indian Agricultural Statistics Research Institute, New Delhi in 1979. He worked as "Research Fellow" for six months with fellowship Rs.1000/- month in "Ford Foundation" New Delhi and completed his Ph.D. degree in 1991 from "Indian School of Mines, Dhanbad" Bihar in "Sampling Theory" with 12 research papers published in single authorship in Journal of Indian Society of Agricultural Statistics, New Delhi. For his continuous works he was awarded with prestigious "Dr. Radhakrishanan Award" on 4th Sept., 1992 by M.P. State U.G.C. and was declared as "The Man of Physical Sciences".

His papers have been referred in various text and reference books published by national and international publishers. He beard the responsibility of chairmanship in "Mathematics & Statistics Session" set up to evaluate the oral presentation of young scientists up to 28 years of age in scientific conference sponsored by MAPCOST, Bhopal and hosted by J.N.K.V.V., during Feb. 27-28, 1993 to give them "Young Scientist Award" at state level. His lecture in workshop On "Frontiers of Research in Finite Population Inferences" during (Nov. 1-9, 1993) financed by Govt. of India secured him second rank.

Dr. Kushwaha is an appointed member of (i) Executive Board and (ii) Editorial Board of "Bulletin Sample Survey Research Group, India", "Academic Council of R.D. University, Jabalpur," to guide Ph.D. Thesis of Scholars. He has published 19 research papers in the journals of National & International repute.

At present, he is working as a Senior Associate Professor (Statistics) in the Department of Mathematics and Statistics, Jawaharlal Nehru Krishi Vishwavidayalaya, Jabalpur, M.P., India.

Dr. Rajesh Kumar graduated from University of Delhi in B.Sc (Honours) in Mathematical Statistics in 1980 and M.Sc. in Mathematical Statistics in 1982. He did Professional Statistician's Certificate Course (P.S.C.C.) in agricultural statistics equivalent to M.Sc. (Agricultural Statistics) in 1983 from Indian Agricultural Statistics Research Institute, New Delhi. He was awarded Gold Medal in Professional Statistician's Certificate Course (P.S.C.C.) in Agricultural Statistics for securing highest marks and V.V.R. Murthy Memorial Prize. He earned Ph.D. degree in Statistics for his work in the field of Survey Sampling in 1995 from University of Lucknow.

Dr. Rajesh Kumar qualified Agricultural Research Services Examination in 1985 and joined Indian Institute of Sugarcane Research, Lucknow as Scientist (Agricultural Statistics). Presently, he is working as Principal Scientist (Agricultural Statistics) at Indian Institute of Sugarcane Research, Lucknow, U.P., India.

He was elected Fellow of Royal Statistical Society, London in 2000 and was granted the status of Chartered Statistician of Royal Statistical Society, London, UK in 2003. He was the elected Secretary of Association of Sugarcane Technologists of India, c/o Indian Institute of Sugarcane Research, Lucknow in 2004 and an elected member of Executive Council of Indian Society of Agricultural Statistics, c/o IASRI, New Delhi in 2006. He was elected as one of the members of International Board of Consulting Editors of Sugar Tech- An International Journal of Sugar Crops and Related Industries published by International Association of Professionals in Sugar and Integrated Technologies, China and Society for Sugar Research and Promotion, India in 2007. He was elected as Editorial Secretary of Bhartia Krishi Anusandhan Samiti, Karnal in 2006.

Dr. Rajesh Kumar has published more than 100 research papers in the field of Statistics, Agricultural Research and Crop Planning. He shared the responsibility in organizing several National Seminars on Sugarcane Research. On many occasions, he was the member of monitoring team constituted for monitoring the trials of All India Coordinated Research Project on Sugarcane conducted at different location in India.